Wind Turbine Technology

Wind Turbine Technology

A.R. Jha, Ph.D.

CRC Press
Taylor & Francis Group
Boca Raton London New York

CRC Press is an imprint of the
Taylor & Francis Group, an **informa** business

CRC Press
Taylor & Francis Group
6000 Broken Sound Parkway NW, Suite 300
Boca Raton, FL 33487-2742

First issued in paperback 2017

ISBN-13: 978-1-4398-1506-9 (hbk)
ISBN-13: 978-1-138-11533-0 (pbk)

Library of Congress Cataloging-in-Publication Data

Jha, A. R.
   Wind turbine technology / A.R. Jha.
      p. cm.
   Includes bibliographical references and index.
   ISBN 978-1-4398-1506-9 (hardcover : alk. paper)
   1. Wind turbines. 2. Wind power. I. Title.

TJ828.J53 2011
621.31'2136--dc22
                                                    2010023856

Visit the Taylor & Francis Web site at
http://www.taylorandfrancis.com

and the CRC Press Web site at
http://www.crcpress.com

This book is dedicated to my beloved parents who always encouraged me to pursue advanced research and development activities in the fields of science and technology for the benefit of mankind.

Contents

Foreword

This book appears at a time when the future well-being and industrial development of nations involved in the 21st century's global economical progress are strictly dependent on the availability of oil and gas at reasonable prices to generate electricity. Deployment of wind turbine technology offers not only a reliable and cost-effective approach, but eliminates the dependence on foreign oil. Furthermore, wind turbines are capable of generating large amounts of electricity at lower costs without generating harmful gases.

Historically, wind energy has been used since 5000 BCE; the earliest use entailed propelling boats. The Sinhalese in Sri Lanka used monsoon winds to power furnaces in 300 BCE. By 200 BCE, the Chinese utilized wind machines for pumping water and the Persians were grinding grains with wind devices. The development of windmill technology in Europe started in the early part of the 12th century. The first windmill to generate electricity was built in Scotland in 1887, followed by the development of large windmill turbines in 1890 in Holland. Larger and more efficient wind turbines were built in the United States during the first half of the 20th century. The level of activity for developing alternate energy sources has been dependent on the cost of fossil fuels; the interest in wind turbines declined with the fall of fossil fuel prices after World War II. However, after the oil embargo of 1973, interest in the development of wind turbines was renewed with the help of the U.S. Federal Wind Energy Program.

The worldwide increases in the demand for and consumption of energy since the latter part of the 20th century may be attributed to the rise of world population and the rapid pace of industrial development in emerging nations such as China, India, and Brazil. It is important to mention that fossil fuels are most economical and therefore they are most widely used. However, they have been the major contributors to environmental deterioration and global climate change. Note that fossil fuels are not in abundant supply. Other energy sources that are currently in use include nuclear power, hydroelectric power, solar energy, and wind. Nuclear power generation offers an attractive alternative energy source, but it suffers from high initial investment cost and dangerous radiation effects. Solar energy-based power generation is an attractive source that offers unlimited output, but it is not available during nights or in heavily clouded environments. Wind turbine technology offers the most attractive renewable energy source because it generates large amounts of electrical energy at lower cost based on unlimited wind energy and inflicts no adverse environmental impacts.

Wind turbines use the kinetic energy of nature to generate electricity for distribution. The wind turbines come in different sizes based on their power generating capabilities. Single small wind turbines with electrical output levels below 25 kW are best suited for remotely located residential homes, telecommunications dishes, and irrigation water pumping applications. A small scale or 50 W generator is sufficient for boat applications. Utility-scale wind turbines with power ratings ranging from 100 kW to 5 MW have proven sufficient to power 100 to 500 homes simultaneously. Many utility company wind farms capable of generating electricity at low generating cost and under maintenance-free conditions have sprung up in rural and onshore regions in recent years. It is of great interest to mention that Denmark fills 40% of its electrical energy needs from wind turbine installations and plans installation of wind turbines capable of delivering more than 4200 MW of electrical power.

Wind turbines can be classified as two major types: horizontal-axis wind turbines (HAWTs) and vertical-axis wind turbines (VAWTs). The advantages of a HAWT include high generating capacity, improved efficiency, variable pitch blade capability, and tall tower base structure to capture large amounts of wind energy. Regardless of classification, a wind turbine converts the kinetic energy of the wind into mechanical power that drives an electrical generator. Appropriate tapering of the rotor blades is selected to maximize the kinetic energy from the wind. Optimum wind turbine performance is possible when a turbine is mounted on a tower of appropriate height in an onshore or offshore region.

This book contains eight chapters, each dedicated to a specific design aspect of wind turbine technology. The book is well organized and emphasizes both theoretical and practical aspects of wind machines. Mathematical equations are provided for important turbine design parameters. I strongly recommend this book to a broad audience including students, design engineers, research scientists, project managers, environmental scientists, and renewal energy source planners. In summary, wind energy sources are bound to play an important role in 21st century economic development; government policy encourages rapid development of renewal energy sources in the United States. The author of this book, Dr. A. R. Jha, presents a comprehensive guide to wind turbine technology and the technical considerations for designing wind turbines for various applications.

Shobha Singh, PhD
Distinguished Member of Technical Staff (Retired)
Bell Laboratories, Murray Hill, New Jersey and
Fellow and Laboratory Director (Retired)
Polaroid Corporation, Cambridge, Massachusetts

Preface

Considerable interest in renewable energy sources and significant increases in cost of foreign oil have compelled various countries to search for low-cost energy sources and technologies such as solar cells, wind turbines, tidal wave turbines, biofuel sources, geothermal technology, and nuclear reactors to achieve lower cost of electricity generation. The future and welfare of Western industrial nations in the 21st century's global economy increasingly depend on the quality and depth of technological innovations that may be commercialized at a rapid pace. Rapid development of low-cost energy sources is not only urgent to reduce the cost per watt of the electricity, but also to eliminate the dependency on and blackmail by oil-producing countries that are hostile toward Western nations, particularly the United States and European countries.

It is important to mention that several countries now deploy wind turbine technology to meet their electrical energy requirements at a lower cost per kilowatt hour.

Rapid design and development activities must be undertaken in the fields of low-cost and efficient micro-hydro turbines and high-capacity wind turbines. Some European and third world countries are constructing wind farms to reduce their electric costs and eliminate dependency on foreign oil. International energy experts predict that wholesale electricity prices will rise 35% to 65% by the year 2015. Under these circumstances, several countries are focusing on deployment of alternate electrical energy sources involving micro- and medium-capacity wind turbines at appropriate installation sites. Studies performed by the author on small- and medium-capacity wind turbines indicate that these wind sources are best suited for land-based defense installations, schools, residential uses, shopping centers, and large commercial buildings. The studies also demonstrate that micro wind turbines may meet the electricity needs of individuals for lights, fans, air conditioners, computers, fax machines, water coolers, microwave ovens, and other electrical devices and accessories.

This book presents a balanced mixture of theory and practical applications. Mathematical expressions and the derivations highlighting their importance are provided for the benefit of students who intend to pursue further studies in wind turbine technology. This book is well organized and covers critical cutting-edge design aspects of this technology. It is written in language appropriate for under-graduate and graduate students who hope to expand their horizons in this innovative field.

This book was written specifically for engineers, research scientists, professors, project engineers, educators, and program managers deeply engaged in the design, development, and research related to wind systems for various applications. It should be most useful to those who wish to add to their knowledge of renewable alternate energy sources. The author has made every attempt to provide well-organized material using conventional nomenclature and consistent sets of symbols and units for rapid comprehension by readers with minimal knowledge of wind turbine technology. The latest performance parameters and experimental data on operational wind turbines provided throughout this book have been taken from various references with due credit given to authors and sources. The bibliographies include significant reference materials. Each of the eight chapters is dedicated to a specific topic relevant to both residential and commercial applications.

Chapter 1 describes the chronological development of wind turbine technology and its generation of electrical energy at lower cost and without greenhouse effects. Wind machines offer lower costs compared to coal-fired thermal plants and nuclear reactor-based power sources. In brief, a wind turbine offers a cost-effective alternative renewable energy source. It is interesting that a wind turbine works as the reverse of an electrical fan: it converts the kinetic energy of the wind into electrical energy. Wind turbines come in different forms and sizes. Small or micro wind turbines are typically rated below 10 kW and are best suited for individual homes, remote operating sites, telecommunication systems, and irrigation applications. Medium-capacity machines are typically rated between 10 and 100 kW. Utility-scale wind turbines have power output ratings ranging from 100 kW to 2.5 MW and are best suited for supplying electricity to local utility companies. When several turbines are installed in wind farms, electricity generation costs approach those realized by coal-fired and nuclear reactor-based power sources.

Several developed and developing countries are leaning toward deployment of micro turbines because these wind power sources are most cost-effective for small business owners in isolated or less densely inhabited regions. Coastal European countries have installed several medium-capacity wind turbines to operate machine shops, irrigation projects, and tube wells. Denmark meets at least 40% of its electrical energy needs from wind turbines installed in coastal regions where average annual wind velocity exceeds 15 mph. China has installed several hundred utility-scale turbines in the Gobi desert where average wind speed is consistently greater than 30 mph.

Chapter 2 deals with various types of wind turbines with major emphasis on design aspects, installation costs, and performance capabilities. It is interesting to mention that the energy produced by a wind turbine is useful in a thermodynamic sense. A wind turbine drives an electrical generator whose output can be stored in batteries or linked to commercial utility power grids. These turbines can be used to produce clean fuels composed of hydrogen and oxygen or to develop fuel cells for other applications. In addition, the kinetic energy generated by a wind turbine can be converted into electrical, chemical, or thermal energy or potential. It is important to note that a wind turbine installed in the right location offers high efficiency and year-round reliable operation with minimum operation and maintenance costs.

Since the early 1950s, several types of high-capacity wind turbines have been manufactured by large companies such as General Electric in the United States and Mitsubishi in Japan. Various types of wind turbines ranging from the earliest versions to modern turbines are shown in Figure 2.1. The windmill has played a key role in water pumping applications. More than one million wind wheels have been used for water pumping in different parts of the world, particularly in remote locations. Wind turbines are classified as two types: the horizontal axis wind turbine (HAWT) and the vertical axis wind turbine (VAWT) as illustrated in Figures 2.2 and 2.3, respectively. Most high-capacity modern wind turbines belong to the HAWT category based on their straightforward design configuration, higher structural integrity, and improved dynamic stability under strong wind conditions. Regardless of the category, high tip speed ratios (TSRs) ranging from 2 to 8 are required to achieve optimum performance in terms of reliability, dynamic stability under violent wind conditions, power coefficient, and power output.

Chapter 3 is dedicated to design aspects and performance capabilities of wind turbine rotors. The chapter commences with a discussion of rotor classification, then moves to critical design requirements and performance capabilities of rotors with emphasis on reliability and dynamic control aspects. Various types of rotors widely used by modern wind turbines and aspects of output performance and design simplicity are described. One-dimensional (1-D), two-dimensional (2-D), and three-dimensional (3-D) aerodynamic analyses of potential wind turbine rotors along with power and thrust coefficient aspects are discussed. Critical circular control flow parameters and their effects on axial velocity components will be described. Note that axial velocity components play key roles in the overall performance and dynamic stability of rotors that operate in various wind environments. Use of the Bernoulli equation is recommended for calculations of performance far upstream, in front of, and just behind a rotor. The axial momentum equation will be derived using certain assumptions. The expression for the rotational speed of the rotor will be developed as a function of axial induction factor and axial wind velocity. TSRs for optimum rotor performance are indentified.

Upwind and downwind rotors are widely used by high-capacity HAWTs while the Savonius and Darrieus rotors are best suited for VAWTs. The efficiency of the Savonius rotor is generally poor over a TSR of 1 to 2.5. Efficiency improves when a TSR exceeds 3. The efficiency further improves to 53% and 57% as a TSR approaches 3 and 4, respectively. The efficiency of a Darrieus rotor is generally poor and varies from 28% to 32% over a TSR ranging from 5 to 7. The maximum efficiency of 33% is achieved at a TSR of 6. Because of poor rotor efficiency, deployment of the Darrieus rotor is restricted to low-capacity wind turbines. The aerodynamic studies of rotor configurations performed by the author indicate that the upwind and downwind rotors demonstrated higher efficiencies and, therefore, are widely used by high-capacity HAWTs. The ideal efficiencies for such rotors range from 51.2% at a TSR of 2 to 57.3% at a TSR of 4 and 58.3% at a TSR of 6.

The principal objectives of Chapter 4 are to apply the basic parameters of the blade theorem and then specify the blade design requirements for given wind turbine systems. The performance of a wind turbine is strictly dependent on the number of blades assigned to the rotor. The blade theory states that the outer contour of the blade must be based on aerodynamic considerations and the materials used in its fabrication must be fully capable of providing the required structural strength and stiffness. Potential models using actual flow conditions to determine the characteristics of airfoils will, be investigated. Expressions for the torque and power coefficients as functions of TSR, drag-to-lift coefficient, and blade radius will be derived. The aerodynamic considerations of the blade area can be modified to ensure greater structural integrity under severe wind environments. Equations for the deflections and bending moments at specific points along the lengths of blades are provided. Effects of loadings on the turbine rotor caused by the earth's gravitational field, inertial loading, and aerodynamic loading on the mechanical integrity of blades will be identified with emphasis on tensile stress, compressive stress, and deflections at the loading and trailing edges of blades. Critical performance parameters for various rotor blades will be discussed.

Chapter 5 describes the sensors and devices required to maintain the dynamic stability and improved performance of rotors under variable wind conditions. Stall regulation, pitch control, and yaw control mechanisms are described and reliability, safe operation, and enhanced turbine performance are emphasized. Control mechanisms using microelectrical mechanical systems (MEMS) and nanotechnology-based sensors will be given serious consideration due to their fast response and higher accuracy for monitoring vital turbine operational parameters. The connection between the rotor blades and hub must be extremely rigid to maintain safe operation under variable wind conditions. Teetering of the hub design reduces the loads on the turbine, thereby yielding higher dynamic stability of a turbine, regardless of electrical load. Critical associated components such as the rotor, generator, transmission system, and other elements responsible for the system reliability, safety, and dynamic stability will be discussed in great detail. A gear box is required to reduce the rotational speed of the rotor needed to maintain the dynamic stability of the turbine under higher tip speeds. As mentioned earlier, most monitoring sensors and safety devices are located at the tops of towers.

Chapter 6 discusses the performance capabilities and limitations of stand-alone wind turbine (SAWT) systems. Structural aspects play a key role in determining SAWT performance capabilities and limitations. SAWTs are best suited for powering remote sites. They can deploy small or moderate capacity wind turbines and banks of batteries to carry electric power for household consumption during winter and summer periods. Studies performed by the author indicate that initial installation cost, relatively low power generating capability, and need for frequent maintenance are the principal drawbacks of SAWT power. However, new SAWT technology has solved some of these problems and, as a result, people in remote areas who cannot access commercial utility power are accepting the SAWT concept. Data compiled by the Pacific Gas and Electric Company indicate that the number of SAWT power systems mushrooms at a rate of 29% annually and that the market for SAWT systems will continue to expand in areas not served currently by commercial utility companies.

Implementation of solar cell technology in a stand-alone wind turbine system offers a cost-effective and practical choice for the customers in remote and inaccessible areas where commercial utility power lines do not exist. Various configurations for SAWT power sources must be investigated in terms of installation cost, maintenance and operational costs, sustainability, and reliable performance.

Chapter 7 deals with wind energy conversion techniques for wind turbine power sources operating in built environments. It is important to mention that the installation of wind turbines in built environments presents complex problems that rarely surface in open environments. Any technique for converting wind energy into electrical energy in built environments must be cost-effective and incur minimum maintenance and operating costs. The selected energy conversion technique must deal with the judicious integration of wind turbine and building aspects in such a way that the building structure concentrates the wind energy exclusively on the turbines. Chapter 7 proposes three basic principles that allow concentration of wind energy at various locations on buildings. The building design determines the optimum locations of wind energy concentrators of four types:

- Wind turbines installed in wind ducts through adjacent buildings
- Wind turbines installed at the tops of the buildings
- Wind turbines installed at the sides of the buildings
- Wind turbines installed between two airfoil-shaped buildings

The aerodynamic aspects of the four distinct concentrators and those of the wind turbines must be properly integrated to produce maximum energy yield. Mathematical models will be used to simulate air flow and verify parameters obtained through computer analysis. Real-time measured data can be obtained in an open jet wind tunnel. Computer simulation can be performed via a computational fluid dynamic (CFD) code.

Chapter 8 focuses on environmental issues and economic factors affecting wind turbine installation. Environmental issues such as physical restrictions, noise levels, tower design constraints, disturbance of local ecology, adverse effects on electronic radio and television signals, and impacts on bird populations must be seriously considered prior to selection of a turbine installation site. Economic factors include the initial cost for analysis, design, fabrication, assembly test, transportation, final checkout testing, installation of turbine and tower, and maintenance and operation. Finally the cost of generation of a kilowatt of electricity must be evaluated prior to the installation of a wind turbine. Economic feasibility is determined strictly on the basis of costs and the value of the electrical energy generated based on preliminary cost estimates.

I wish to thank Dr. Ashok Sinha (retired senior vice president, Applied Materials, Inc., Santa Clara, California), for his critical review and also for accommodating last-minute additions and changes to retain consistency of the text. His suggestions helped the author prepare a manuscript with remarkable coherency. Last, but not least, I wish to thank my wife Urmila Jha, my daughters Sarita and Vineeta, my son-in-law Anu, and my son Lt. Sanjay Jha, who inspired me to complete the book on time under a tight schedule. Finally, I am very grateful to my wife who was very patient and supportive throughout the preparation of this book.

A. R. Jha

Chapter 1

Chronological History of Wind Turbine Technology

1.1 Introduction

This chapter describes the history of wind turbine technology development and its potential applications. Wind turbine technology offers cost-effective solutions to eliminate the dependence on costly foreign oil and gas now used to generate electricity. Additionally, this technology provides electrical energy without greenhouse effects or deadly pollution releases. Furthermore, wind turbine installation and electricity generating costs are lower compared to other electrical energy generation schemes involving coal fired steam turbo-alternators, tidal wave turbines, geothermal-, hydrothermal-, biofuel-, and biodiesel-based electrical energy sources and nuclear reactor-based generators.

Wind turbine technology offers a cost-effective alternate renewal energy source. It is important to mention that a wind turbine is capable of generating greater amounts of electrical energy with zero greenhouse effects compared to other energy generating schemes including solar cell, tidal wave, biofuel, hydrogen, biodiesel, and biomass technologies. A wind turbine is the reverse of an electrical fan. A wind turbine uses wind energy to generate the electricity; a fan uses electricity to generate wind. In more sophisticated terminology, a wind turbine converts the kinetic energy of the wind into electrical energy.

Wind turbines come in different sizes and types, depending on power generating capacity and the rotor design deployed. Small wind turbines with output capacities below 10 kW are used primarily for residences, telecommunications dishes, and irrigation water pumping applications. A wind turbine advertisement states that a prototype 5-kW system can be built for $200, using inexpensive off-the shelf components and excluding labor costs. Utility-scale wind turbines have high power ratings ranging from 100 kW to 5 MW—sufficient to power 10 to 500 homes.

Current wind farms with large capacity wind turbine installations have sprung up in rural areas and offshore regions and are capable of generating electricity in excess of 500M MW for utility companies; they present much lower generating costs and zero maintenance and operating costs. Denmark developed wind turbine technology as early as 1970 and expects to have wind turbine capacity exceeding 4500 MW by the end of 2010. Furthermore, Denmark now obtains at least 40% of its electrical energy from wind turbine installations, thereby realizing significant reductions in electricity generation costs, greenhouse gas emissions, and atmospheric pollution.

Modern wind turbines are classified into two configurations: horizontal-axis wind turbines (HAWTs) and vertical-axis wind turbines (VAWTs), depending on rotor operating principles. The VAWT configuration is similar to an eggbeater design and employs the Darrieus model named for the famous French inventor. Regardless of classification, a wind turbine converts the kinetic energy of the wind into mechanical power that drives an alternating current (AC) induction generator to produce electricity.

HAWTs with two or three blades are the most common. Wind blowing over the propeller blades causes the blades to "lift" and rotate at low speeds. Wind turbines using three blades are operated "upwind" with rotor blades facing into the wind. The tapering of rotor blades is selected to maximize the kinetic energy from the wind. Optimum wind turbine performance is strictly dependent on blade taper angle and the installation height of the turbine on the tower. General Electric Company pioneered the development of wind turbine technology by exploiting its unique capabilities such as improved performance under variable wind environments, variable speed control under reduced load conditions, and cost-effective electronics for local grids. More than 5,000 General Electric wind turbines generating 1.5 MW are in use worldwide because of their high reliability and efficiency.

The major components of a wind turbine include a low-speed rotor consisting of two or three light-weight blades with optimum airfoil shapes operating at 30 to 60 RPM, a high-speed shaft mechanically coupled to low-speed via a gear box assembly and operating between 100 and 200 RPM, a pitch motor drive assembly, a yaw motor drive assembly, a nacelle, a wind vane indicator, an AC induction generator operating at high speed, a speed controller unit, a tower structure, an anemometer, and other accessories necessary to provide mechanical integrity under heavy wind gusts. Winds are generally caused by uneven heating of the atmosphere by the sun, the irregularities of the earth's surface, and the rotation of the earth. However, wind flow patterns are modified by the earth's terrain features, bodies of water, and surrounding vegetation. The power generated by a wind turbine can be used to perform specific tasks such as grinding grain, pumping water, or driving an AC induction motor to produce electricity that can be fed into a utility grid system. In addition to the economic and environmental benefits provided by wind turbines, the technology eliminates the dependence on costly foreign oil and neutralizes the political blackmail tactics used by hostile oil producing countries.

1.2 Major Benefits and Problems Associated with Alternate Energy Sources

The United States is the world's greatest consumer of electricity. According to published data on energy sources, the U.S. gets about 38.1% of its energy from petroleum, 22.9% from natural gas, 23.2% from coal, 8.1% from nuclear plants, 2.9% from biomass, 2.7% from hydroelectric sources, 1.7% from propane, 0.3% from geothermal sources, 0.15% from wind turbines, and 0.1% from solar cells. Based on the published data, the worldwide electricity generation is roughly 1% from wind turbines. However, several countries are considering installations of wind turbines because of the high costs of fossil fuels and unpredictable petroleum supplies from OPEC countries. As far as the cost of electricity generation (cents per kilowatt hour) is concerned, electricity generated by wind turbines is cost-competitive with electricity from conventional energy sources such as coal fired power plants.

Renewable energy sources present many benefits. They offer clean, uninterrupted, environmental impact-free electrical energy at reasonable cost. Studies performed by the author indicate that wind and solar sources offer the cleanest and most cost-effective renewable energy. The studies further indicate that wind turbine technology is best suited for desalinating seawater, crop irrigation, and food production in coastal regions. Wind turbines are effective for desalination of water in coastal areas that lack fresh water.

Any energy generating source or scheme requires correct installation and specific operating conditions to ensure cost-effective operation. Some sources require large initial investments during the installation phase, while others require unique operating environments available only in a specific geographic location or terrain. As stated earlier, every energy source (coal-fired generation, nuclear power, hydraulic turbines, tidal wave turbines, geothermal and hydrothermal energy sources, biofuel, conventional diesel generators, and wind turbines) suffers from certain drawbacks. Coal-fired power generating sources produce harmful polluting gases and aerosols. Nuclear power plants require heavy initial investment and present radiation dangers. Hydro-turbine installations require water reservoirs to ensure constant water pressure, and tidal wave turbines require ocean waves with specific characteristics available primarily in Portuguese coastal regions. Geothermal and hydrothermal energy sources require low- and high-temperature underground regions to harness energy sources; such locations are known to exist only in Iceland. Biodiesel power sources require large supplies of raw materials.

Note that Iceland meets 70% of its heating and electricity needs utilizing geothermal and hydrothermal energy sources in areas of low and high temperatures. Energy experts in Iceland predict that by 2025 more than 80% of its electric energy requirements will be satisfied by clean energy from geothermal and hydrothermal sources. The geothermal and hydrothermal sources of Iceland arise from the presence of two tectonic plates of the Mid-Atlantic Ridge System. These two plates face each other but move in opposite directions. The action generates enormous churning actions that produce high-temperature regions underneath the ground. To accelerate the development of Iceland's geothermal and hydrothermal energy sources, a deep drilling project established in 2008 is capable of drilling 5 km deep to bring bore holes into contact with water at temperatures ranging from 400 to 600°C. This high-temperature water can be used for heating buildings or producing superheated steam to drive large turbo-alternators to generate 2000 to 4000 MW of electricity. Currently, the project focuses on well depths that will reach into hydrothermal vents most ideal for heating. Iceland expects to capture as much as ten times the current geothermal power in near future. It has allocated limited sources to develop wind turbine farms to generate electricity.

It is important to mention that despite manageable shortcomings or problems, solar and wind turbine energy sources offer electrical energy with lower installation costs, higher reliability, and cost-effective operation. These two alternate energy sources provide clean energy free of environmental effects along with minimum cost and complexity compared to other energy sources mentioned above. Preliminary studies performed by the author predict that wind turbine technology will able to meet 10% of the world's electricity requirements by 2020.

Hybrid power generating schemes are occasionally found most reliable and cost-effective. A combination of wind turbine technology with a hydroelectric power generating source would be attractive and practical because it takes very little time and effort to open or close a valve to a water turbine. Note that water stored in a reservoir can be used to drive a water turbine when the operation of a wind turbine is not feasible, for example, during strong winds.

1.3 Benefits and Disadvantages of Wind Turbine Technology

1.3.1 Benefits

Compared to other energy sources wind turbine technology offers affordability, pollution-free and maintenance-free operation. Major benefits of wind turbine technology can be briefly summarized as follows:

- It saves substantial money on utility bills; users face no power shortages or failures as experienced by customers who depend on electrical utility grids.
- It delivers environmentally friendly and efficient electrical energy at lower cost, particularly, in areas where electrical grids are not available, for example in remote locations with difficult terrain features.
- Installation does not jeopardize the value of a home, office building, or commercial building. The installation can be easily undone and leaves no adverse visible effects at installation sites.
- The turbine does not require frequent or intermittent maintenance or employment of operations personnel; unlike steam and gas turbine-based alternator systems, no maintenance or operational costs are incurred.
- The technology essentially offers home-made electrical energy and off-grid living, which is not readily possible with other technologies.

Despite these benefits, wind turbine technology has a few shortcomings. Damage to its tower structure or housing caused by strong winds may necessitate costly repairs or maintenance, particularly if a component must be replaced on a tower typically ranging from 50 to 70 m in height.

1.3.2 Disadvantages

It would be unfair not to identify the potential disadvantages or the imminent dangers. Consistent noise is the most annoying factor. For example, a constant high wind noise problem may be overcome by selecting installation sites away from residential areas, schools, and commercial buildings. It is important to understand that wind turbines are generally loud and the turbine blades represent a danger to birds flying less than 350 ft above the ground. Older turbines built 15 to 25 years ago are not only louder, but also less cost-effective. Turbine noise increases with operating height, rotor blade size, and power output. Most current wind turbine blades vary in length from 50 to 150 ft and the heights vary from100 to 300 ft, depending on ground surface roughness, vertical distribution of wind speeds, local terrain features, number of turbines on wind farm, and electrical power generating requirements.

Wind turbine installation cost depends on site selection, tower height, and output power rating. A typical 1.5-MW General Electric wind turbine costs around $2.5 million including parts, installation, and site selection. This cost will be much higher for wind turbines with higher output ratings. Other costly factors are site selection and tower structure. However, the cost of producing wind power has dropped four-fold since 1980, based on Electric Power Research Institute estimates.

A wind turbine normally sits about 30 stories above the ground at the hub, where three 130-ft blades are connected to the tower head. Although maintenance is rarely required after careful installation, an ordinary maintenance or erection technician is not qualified to perform the required tasks. Turbine installation and maintenance require intelligence, quick thinking, split-second decision making, stamina, strong knees, and the ability to function at dizzying heights. In other words, a technician must be hypervigilant while working in tight spaces at heights exceeding 235 ft, in the presence of high-voltage electrical cables and spinning metal components. A turbine tower contains no elevator. Access to the top is only possible by climbing rung by rung on a narrow-steel ladder inside the tower. It is extremely difficult for an ordinary technician to perform tasks in such a narrow environment. However, an agile and experienced technician will be able to reach the tower top in 10 minutes or less several times a day. Wind turbine technicians require 8 weeks of rigorous training and classroom instruction covering installation, operation, and maintenance. Under working conditions, wind turbine technicians face inherent dangers while working alone almost 300 ft in the air with little support. However, crack wind turbine technicians earn annual salaries in six figures.

1.3.3 Unique Installation Requirements

Wind turbines require unique installation specifications to operate with optimum efficiency. These strict installation requirements must not be characterized as major disadvantages. For example, wind turbine installers and designers carefully consider turbine sites and wind farm conditions. They must select optimum parameters for rotor blades, spacing between turbine units, height of the turbine and tower structures, and other conditions that will yield reliable, safe, and efficient operation with minimum maintenance and operation costs. The visual impact of a wind turbine is viewed as a negative factor because of the large blade spans—about 25 m in the 1980s and as large as 125 m today. The new turbine rotors developed in Germany ride even higher. They are installed on towers as high as 100 m or 328 ft. Hub heights generally "maxed out" at 70 m in the 1990s.

1.3.4 Repowering to Increase Output by Existing Turbines

Higher hub heights and re-powering schemes are highly desirable, if optimum turbine performance and maximum profit are the principal requirements. It is important to mention that higher hubs place the wind turbines in stronger and more consistent winds, thereby increasing the number of operational hours per year that a wind turbine will operate most efficiently. Performance data from the Denmark Wind Turbine program indicate that wind farms with double the electrical power generating capacity of the original farms can deliver as much as four times more electrical energy with no changes in turbine design parameters. A repowering scheme is intended to harness more electrical energy by replacing existing wind turbines with higher capacity units. Based on height restrictions, a repowering wind turbine site can generate 50% more electrical energy. A wind turbine farm must achieve its full potential to meet the cost-effective design criteria.

The German government is planning to approve a wind-farm repowering scheme that will produce more electricity by replacing several small wind turbines with a few large units [1]. Several current wind farms will be replaced with farms of larger turbine units that will significantly increase the electrical energy produced by the farms. This trend is spreading in other countries because it achieves significant reduction in the cost of generating electricity with wind turbine technology.

1.4 Worldwide Utilization of Wind Turbines

Several countries are looking into alternate methods of producing clean energy that will eliminate dependence on costly foreign oil and gas. Rising costs for foreign oil and gas have ruined the economies of several countries. Since wind turbines are capable of generating large amounts of clean electrical energy ranging from 50 to 500 MW with minimum initial investment, the wind turbine technology is commanding great attention. The technology is fully matured and turbines with various output capacities are available to meet residential, commercial, and industrial requirements. Wind turbine operation activities and installations in various countries will be described, with particular attention paid to unit output capacity and unique performance parameters.

Today's electricity generation windmills can reach heights close to that of the Washington Monument, with blade lengths exceeding 100 m or 328 ft. Current wind turbines are rated up to 5 MW generating capacity, which will be able to meet energy requirements for hundreds of homes. Wind turbine technology mature; the machines are reliable and operate continuously when constant wind is available. In April 2008, a small town in Missouri with a population close to 1500 became the first city in the United States to obtain all its electricity from wind power provided by a 5-MW, four-wind turbine farm.

1.4.1 Denmark

Denmark pioneered wind energy technology as early as 1970. Installations widely known as wind farms were built in large numbers to generate large amounts of electrical energy. The country is so bullish that it has plans to convert a 77-wind turbine facility into a 13-turbine facility using devices with larger capacities. Each large capacity turbine will be able to generate four times more compared to existing 2.5-MW capacity turbines.

According to the Historical Wind Turbine Development (HWTD) survey, a 50-kW wind turbine facility went into operation as early as 1941 and a 100-kW operation went online in 1957 in Denmark. Wind turbines with higher megawatt capacities were installed between 1980 and 1990 in coastal regions. Denmark plans to generate more than 4500 MW of electrical energy by the year 2020 using wind turbine technology. Its parliament is planning to introduce legislation in 2009 that will require wind park operators to compensate residents if installations of wind turbines reduce their property values. The legislation will also authorize the operators to dismantle the old and inefficient wind turbines and sell or ship them to Eastern European countries. All these activities indicate that the country is about to expand the deployment of wind turbine technology to meet the needs of its citizens.

1.4.2 Germany

Scientists and engineers in Germany developed wind turbine technology to generate electricity early but delayed the implementation of the technology to generate large amounts of electrical energy for consumers. Several wind turbine installations with kilowatt ratings started to operate in the early 1980s in coastal regions where wind gusts were frequent. According to the HWTD survey, German engineers installed a 10-kW wind turbine facility as early as 1936 and a 30-kW facility in 1958 and added more high-capacity wind turbine installations between 1980 and 2000. According to the Wind Energy Industry Association of Germany, the country plans to add at least 15,000 MW of new wind-power capacity by 2020. The association selected new turbine sites in hilly locations and coastal regions where optimum wind energy levels are constant throughout the year. According the latest news on wind turbine-based energy generation sources, the German government plans to approve a large wind farm scheme that will produce more electricity with many small-capacity wind turbines [1].

1.4.3 China

The Peoples' Republic of China has large tracts of open lands in the Gobi Desert region, where high wind speeds are prevalent throughout the year. The Gobi Desert is the most ideal location for wind turbine installations and most wind turbine facilities have been installed in that region. Between 2000 and 2008, China installed hundreds of 100-MW wind turbines in the Gobi Desert area and plans to install more wind turbines in some regions of Tibet, where open land with year-round high winds is readily available. Chinese energy experts predict that plans to install more high-capacity wind turbines in the Gobi region will boost the current power generation capability to exceed 10,000 MW by the 2012. Most of these wind turbine generators are connected to utility power grid systems to ensure continuous supplies of electrical energy at minimum cost to consumers. Since China's population is approaching 1.3 billion people, electricity generated by wind farms will be more affordable than electrical energy produced by coal-fired steam turbo-alternators or nuclear power reactors.

The availability of large quantities of electricity at minimum cost is critical because of China's huge population. Deployment of wind turbine technology offers clean and economical electricity; available wind energy is unlimited. Vast open lands at high altitudes in Tibet and open areas in the Gobi desert are ideal sites for generating large amounts of electricity with simple machinery at minimum cost.

1.4.4 United States

The United States put its first 500-kW wind turbine into operation as early as 1987. Several wind farms with capacities ranging from 10 KW to 5 MW were installed between 1995 and 2000. In addition, wind turbines with power generating capacity ranging from 50 to 500 MW were installed between 2000 and 2008. Most of these installations are located in the western coastal regions (California, Oregon, and Washington) and in eastern coastal states (Massachusetts, Vermont, and Maine). Alaska coastal regions are also considered appropriate for wind turbine installations because of strong wind conditions throughout the year. Large flat lands in Montana, Kansas, and similar regions are considered ideal for wind turbine installations; highly reliable wind turbine plants with capacities ranging from 100 kW to 5 MW presently operate at minimum cost. Wind turbines for residential applications operate in North Carolina, New York, Appalachian Mountain states, Maine, Montana, Connecticut, and other areas where high wind speeds are prevalent.

Renewable energy experts note that U.S. wind turbine generating capacity exceeded 18,000 MW at the end of 2005 and surpassed Germany as the top wind power produce, with more than 25,000 MW in place at the end of 2008—enough to provide electricity to 5 million average households. Wind accounted for nearly a third of all new power-producing capacity added in 2008. Deployment of several large windmills is needed to generate electrical energy exceeding 1000 MW and more large-scale wind turbine projects are in the planning stage. A 4000-MW, 400,000-acre wind farm is proposed for the Texas panhandle by the oilman T. Boone Pickens. Energy experts predict that to meet total U.S. energy requirements, wind farms covering approximately 750,000 square miles—roughly equal to the size of Texas, California, Montana, and Florida combined—would be needed. U.S. energy planners predict that wind energy sources could supply 20% of the country's electricity needs by 2030, up from the current 10% electrical energy generation from wind turbine technology. California, Texas, and Iowa are considered the most attractive sites for wind farms capable of generating large amounts of electricity that could be fed into utility power grids.

Although wind energy consumption in the U.S. represents less than 0.5% of the total energy generated, various types of wind turbines manufactured by several U.S. companies to meet the wind energy demands of citizens. Several U.S. manufacturers, dealers, and installers specialize in small-scale wind turbines (10 to 100 kW), utility-scale wind turbines (1 to 5 MW), and large turbines for wind farms (50 to 500 MW).

1.4.5 Canada

Both the eastern and western coastal regions of Canada are well suited for wind turbine installation because of high winds and a few selected areas in the Northern Canada are equally attractive for wind turbine installation. However, access to these regions is extremely difficult due to harsh climatic environments. Despite the variety of ideal locations for wind turbine installations, no capacity wind turbine was installed in Canada until 1987. In the early 1990s, a few low-capacity utility-scale wind turbines with outputs ranging from 5 to 10 MW were installed on eastern coastal areas. Some Small-scale wind turbines with outputs ranging from 10 to 100 kW for residential applications were installed from 2005 to 2008 to provide electricity for lighting and pumping. According to the international energy survey, barely 0.1% of Canada's electricity is generated by wind turbine technology.

1.4.6 Belgium and the Netherlands

Belgium's coastal regions offer high winds throughout the year. Hundreds of wind turbine installations ranging from small-scale (10 to 100 kW) to utility-scale (2 to 5 MW) were installed between 1950 and 2008 to meet electrical energy requirements. The first utility-scale three-blade wind turbine with an output capacity exceeding 300 kW was installed in 1987. According to the electric energy consumption survey, several hundred small-scale facilities with capacities ranging from 10 to 50 kW were installed between 1990 and 2005 for residential and pumping applications, particularly during the flood seasons common to the Netherlands. These small-scale installations are largely visible near dairy farms and open lands.

1.4.7 United Kingdom

Coastal regions of Scotland, England's western coastal regions, and English Channel locations are considered most ideal for installation of wind turbines due to high winds. Wind turbine installations received great attention in England after 1950. The first British wind turbine with a 100-kW power rating was installed in 1955. A second unit with the same rating went into operation in 1957 and was immediately followed by a wind turbine capable of generating electrical energy in excess of 130 kW. The British government put into operation a wind turbine with output capability exceeding 250 kW around 1983. Since then, several installations with higher power ratings have been commissioned in various parts of the country, mostly in the coastal regions where high winds are prevalent. British energy planners in 2008 recommended a project involving 65 wind turbines on nearly 5,000 acres of hillside near Llanbrynmair in Powys, Wales. The project is expected to generate about 400 MW due to deployment of the newest high-efficiency wind turbines. A few years ago, the United Kingdom announced plans to build the tallest wind turbines to generate large amounts of electrical energy.

1.4.8 France

France's eastern and northern coastal regions are best suited for wind turbines. France implemented its first low-power-rating wind turbine (15 kW output) as early as 1929. That was followed by a wind turbine energy system with a power rating of 800 kW in 1958. The French energy agency installed a 1000-kW (1 MW) wind turbine in 1963 to demonstrate the design and development capabilities of wind turbines with higher power output levels. From 1960 to 2008, several wind turbines with power ratings ranging from 1 to 5 MW went into operation to minimize dependence on costly foreign oil. The maximum capacity wind turbines became operational between 1973 and 1977, when oil producing nations of the Middle East severely cut oil production and increased barrel prices ten times over a very short period. France is slowly but consistently adding more wind turbine installations in areas where high winds are prevalent.

1.4.9 Russia

The first wind turbine with power output capability of 100 kW was installed in what was known as the Soviet Union as early as 1931. Deployment of wind turbine technology was not pursued more aggressively because of the availability of abundant oil at minimum cost in the southern states. According to published reports, wind turbines with capacities ranging from 100 kW to 5 MW operate in the northern and southern regions. No reliable information on the number of wind turbine installations and wind farms in regular operation is available. Russia continues to use oil-fired power plants to generate large amounts of electricity.

1.4.10 Italy

Italy is blessed with several coastal regions that are well suited for wind turbine installations and wind farm development. According to various European energy publications, a few wind turbines with outputs ranging from 200 to 500 kW went into operation in Italy between 1975 and 1990. Because unlimited cheap and high-quality oil was available from Libya, the Italian government was not interested in installing larger numbers of wind turbines. Published data on wind turbine installations in Italy indicate that several wind turbines rated at 200 and 225 kW were put into operation after oil prices suddenly increased ten-fold in 1973. Energy experts still believe Italy will not able to install large wind farms because of large populations concentrated in coastal regions that are most suitable for wind energy installations.

1.4.11 Early Wind Turbine Development: Summary

The author wishes to present a chronological development summary of wind turbines commencing from 1800. Published literature on renewal energy sources reveals that Denmark was the first country to install a 20-kW wind turbine as early as 1891 and it added four more wind turbines between 1941 and 1957. Denmark plans to add more wind turbine installations to meet 40% of its energy requirements by 2020. Table 1.1 summarizes data on wind turbine installations in various countries prior to 1990.

Table 1.1 Operating Details of Wind Turbines Installed before 1990

Country	Installation Year	Rated Power (kW)	Diameter (m)	Swept Area (m²)
Denmark	1891	18	23	408
Denmark	1941	50	17	237
Denmark	1942	70	24	452
Denmark	1943	30	18	254
Denmark	1957	200	24	452
France	1929	15	20	314
France	1958	800	30	716
France	1963	1000	35	962
Germany	1926	30	20	314
Germany	1956	200	33	855
Germany	1958	100	34	908
Italy	1981	225	32	804
Italy	1976	200	33	855
Soviet Union	1931	100	30	707
Netherlands	1987	300	30	707
United Kingdom	1955	100	15	177
United Kingdom	1957	100	24	468
United Kingdom	1965	130	25	491
United Kingdom	1983	250	20	314
United States	1987	500	34	955

Note: Data indicate blade diameters and swept area requirements as functions of electrical energy generated. Power generated depends on blade diameter, blade angle, wind direction, tower height, wind velocity, and terrain features.

1.5 Operating Principles of Wind Turbines

It is helpful for readers to understand the basic operating principles of a wind turbine. The operation is based on the scientific theory of fluid mechanics and some elements of aerodynamics. Modern wind turbines catch the wind by turning into or away from air flows. Wind moves the propellers mounted on a rotor and the movement turns a high-speed shaft coupled to an electric or induction generator. Design configurations for a utility-scale wind turbine and small-scale wind turbine (for residential use) are depicted in Figure 1.1. Air currents that blow over flat terrains or over-the-hill regions have wind velocities ranging from 10 to 65 miles per hour and are sufficient to turn large blades attached to a high-speed shaft coupled to a generator [2]. Rotor blade design parameters such as chord, twist angle, and length are selected to achieve optimum aerodynamic performance and acceleration

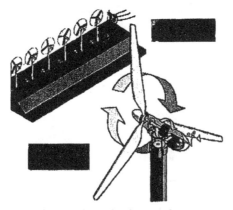

(a) Utility-scale wind turbine configuration

(b) Small-scale wind turbine configuration

Figure 1.1   Wind turbine configuration (a) utility-scale and (b) small-scale.

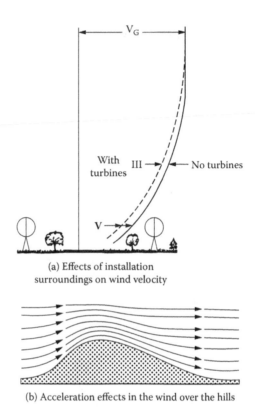

(a) Effects of installation
surroundings on wind velocity

(b) Acceleration effects in the wind over the hills

Figure 1.2 Effects on wind turbine performance from (a) installation ground-based surroundings and (b) acceleration effects over the hills.

effects under various wind conditions at different tower height levels as indicated in Figure 1.2.

1.5.1 Critical Elements of Wind Turbine Systems and Their Principal Functions

As stated earlier, wind turbine technology offers an attractive approach for an alternate source for producing large amounts of electrical energy at minimum cost. It is appropriate to identify the critical elements of a wind turbine and the operating functions provided by the critical components. The critical components of a wind turbine are the tower, rotor blades, anemometer, controller, main enclosure, gear box, low-speed shaft, high-speed shaft, brake mechanism, electrical generator, yaw bearing, yaw motor, yaw drive, pitch indicator, wind indicator, and nacelle and low-loss cables to carry the electrical energy down the tower. A step-up transformer at the base of the tower allows transfer of the wind-generated electricity to the utility

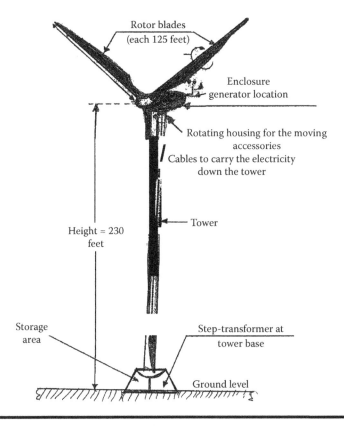

Figure 1.3 Wind turbine components.

power grid. All elements except the step-up transformer are located at the top of the tower as shown in Figure 1.3. The enclosure depicted in the figure rotates to enable the rotor blades to face into or away from the wind. This essentially is the operating principle of a wind turbine. The wind moves the propeller that turns the low-speed and high-speed shafts. The high-speed shaft is connected to a generator capable of producing electrical energy.

The anemometer is a critical element of a wind turbine. It gauges wind speed and direction and sends the information to the controller that in turn provides necessary data to critical elements of the system. The controller essentially directs the yaw motor to turn the rotor to face toward or away from the wind, depending on wind direction. The gear box, the heaviest element of the system, converts the slow rotation (revolutions per minute or RPM) of the low-speed rotor shaft to higher RPM of the high-speed shaft which is mechanically coupled to a generator that produces the electricity. In brief, the high-speed shaft drives a generator that converts mechanical energy into electrical energy. Just behind the hub is the housing (see Figure 1.3) for the gear box and other components.

1.6 Wind Turbine Classifications

The three major categories based on energy generating capacity are utility-scale, small-scale, and intermediate-scale wind turbines and all are commercially available from various sources. Wind turbines are manufactured in various countries including the United States, Denmark, Switzerland, India, United Kingdom, China, Germany, France, Italy, and other technologically advanced countries.

1.6.1 Utility-Scale Wind Turbine Sources

These types of wind turbines are available from various manufacturers and dealers such as General Electric (GE) Wind Energy Source, Mitsubishi Power Systems, Siemens Corporation, Suzion Power Company, Clipper Wind Power, Nordex Corporation, Vestas, and others. The utility-scale turbines manufactured by GE are more reliable and safer according to the company's program manager; its turbines are rated at 1.5, 2.5, and 3.6 MW. According to the program manager, the 2.5-MW wind turbine is the most cost-effective and occupies minimum space. The 1.5-MW GE wind turbine costs around $2.5 million excluding costs for installation and accessories needed to provide additional mechanical integrity for operation in harsh environments.

Mitsubishi Power Systems is another major supplier. Power ratings of its wind turbines range from 2.0 to 2.8 MW. The latest 2.4-MW wind turbine with a 92-m rotor is called the system with "smart yaw" capability because it can be assembled by small cranes at a desired site, can operate with reverse configuration during high winds and under rapidly variable wind directions, and exhibits enormous resistance to wind storms, lightning, and thunder. It is important to mention that yaw components (motor and controller) play a critical role in turbine performance under uncontrollable and unpredictable wind environments. In brief, the dynamic stability of a wind turbine depends totally on the performance capability of the yaw mechanism. Mitsubishi wind turbines operate worldwide and no accidents have been reported to date. Reliability, safety, and improved performance are given serious consideration during the design and development phases. Procurement and installation costs are available from authorized Mitsubishi suppliers.

Siemens Corporation has been a front player in the design and development of wind turbines in European countries. It has manufacturing facilities in Germany and Denmark. Its latest and most popular model is the 2.3-MW wind turbine now used worldwide. The corporate technical department is located in Germany; turbine blades are manufactured in a coastal town in Denmark. The turbine nacelles are manufactured in Brande, Denmark. Siemens recently installed seventy 2.3-MW wind turbines in Texas to provide output sufficient to meet the electricity needs of more than 50,000 households—about 310 households per megawatt of electricity generated by these turbines. Turbine cost is based on power generating capacity. Procurement and installation costs can be obtained from Siemens dealers.

1.6.2 Pay-Off Periods for Utility-Scale Wind Turbines and Wind Farms

Pay-off period is the most important issue in a customer's mind before he signs a contract for installation of a wind turbine. The customer deserves a clear and reliable response from a dealer and an installer. The pay-off period is strictly a function of the costs of turbine procurement, tower procurement, tower installation, delivery costs, and final operation check-out. Based on procurement data from GE, a 1.5-MW wind turbine cost around $ 2.5 million as of January 2008, including the turbine delivery and installation at the site. However, tower procurement and installation costs are not included.

Preliminary estimates on a wind farm installation composed of five 1.5-MW turbines indicate that a wind farm will cost around $12.5 million and will be adequate to meet electricity needs for approximately 2500 homes based on the ability of a 1.5-MW unit to meet electricity requirements of 500 homes. According to the best GE estimates, it would take 10 to 12 years for a 1.5-MW wind turbine to pay for itself, assuming a $500 per year residential electricity bill (equivalent to about $42 per month). Linear interpolation indicates that a $1000 annual electricity bill will reduce the pay-off period to around 5 years. It is important to note that pay-off periods for wind turbines are expected to be less than 50% of the pay-off periods for solar panel installations. However, the pay-off periods for wind farms containing hundreds of utility-scale turbines will be much longer because of the large number of towers and generating units required, as illustrated in Figure 1.4.

Figure 1.4   Large wind farm containing hundreds of turbines.

1.6.3 Small-Scale Wind Turbines

Several companies in the United States, Europe, and China maintain manufacturing facilities for small-scale wind turbines. The procurement and installation costs for high-capacity wind turbines cannot be justified for users whose electricity needs are less than 20 kW. Suzion, a Swiss company, is the major supplier of small-scale wind turbines and has successfully installed small wind turbines in several developing countries that are very pleased with the efficient performance and low procurement costs. The company manufactures small turbines with power ratings ranging from 1 kW to 20 kW. A 1-kW unit costs about $1561; a 2-kW unit, about $2753; a 3-kW unit, about $5449; a 5-kW unit, about $8104; a 10-kW unit, about $13,298; and a 20-kW unit, about $24,410 (prices exclude installation and shipping costs).

Small-scale wind turbine customers are impressed with the performance, efficiency, reliability, aerodynamic design, affordability, and pay-back periods of these small wind turbines. Suzion turbines operate in Europe, India, Singapore, Japan, South Africa, and other countries. They can be purchased directly from manufacturers or authorized dealers in the United States and European countries. In the U.S., dealers include Aerostar Wind Turbines, All Small Wind Turbines, Wind Turbine Industries, Southwest Wind Power, Lake Michigan Wind and Sun, Bornay, Bergey Wind Power, Scoraig Wind Electric, and Appalachian Wind Systems. Table 1.2 summarizes details of small-scale wind turbines manufactured by U.S. companies.

Small-scale wind turbines manufactured by Southwest Wind power are compact, inexpensive, quiet, and highly reliable. They are best suited for residential, remote, telecommunications transmitter, offshore platform, sailboat, and water pumping applications. Southwest has produced more than 90,000 wind turbine generators in the past 20 years.

Wind turbines made by Aerostar deploy induction generators that allow simple direct utility interconnection without need for an inverter. This eliminates reliability problems and decreases system cost. Furthermore, Aerostar wind turbines use the two-blade teetering rotor design that substantially reduces loads on the turbine and tower structure, resulting in a low-cost, efficient, and smoothly running system. Newer versions of wind turbines incorporate advanced materials, technologies, and design concepts such as variable generator excitation control techniques that substantially improve energy capture, particularly at low wind speeds. Other features are aerodynamic rotor control to ensure safe operation and acoustic isolation of moving parts to suppress noise. Aerostar systems are available with a color touch screen displays, data logging of wind speeds and turbine output levels, and Ethernet interfaces for continuous monitoring of performance from anywhere in the world.

Table 1.2 Technical Data for Small-Scale Wind Turbines Manufactured in United States

Manufacturer	Capacity (kW)	Model Number	Technical Data
Abundant Renewable Energy	2.5 and 10.5	E110 and E442	None
Aerostar	10	None	Blade radius = 6 m
AeroVironment	1	AVX-1000	Two-blade teetering rotor
Bergey Windpower	1 and 10	XL.1 Excel	Reliable and Rugged
Gala-Wind Ltd.	11	None	Compact and reliable
Proven Energy Ltd.	2.5 and 6	None	Most stable and reliable
Southwest Windpower	1 and 3	W-200 and W-500	Less expensive
Wind Energy Solutions	2.5	WES-5	Rotor diameter = 5 m
Wind Turbine Industries	10	23-10 Jacobs	Rotor diameter = 7 m

1.6.4 Component Cost Estimates for Small Wind Turbine Systems

Exact cost estimates for wind turbine major components are not readily available. The costs for towers and turbines vary from supplier to supplier. Furthermore, some dealers quote tower cost without installation and delivery costs. Tower cost is also dependent on the height of the structure and the structural integrity requirements. The wind turbine cost is based strictly on the output capacity, rotor design complexity, number of rotor blades, type and location of generator, and complexity of design. The author collected some cost data on towers and small wind turbines with output capacities not exceeding 10 kW. These cost data are summarized in Table 1.3.

1.6.5 Total Installed Cost for 5-kW Wind Turbine in 2009

The total installed cost includes the tower, turbine, installation, delivery, and required steel piping. A break-down of the total cost of a 5-kW small wind turbine are listed in Table 1.4.

Table 1.3   Retail Cost Estimates for Wind Turbines and Associated Towers

Power Output (kW)	Retail Turbine Cost ($)	Retail Tower Cost ($)/ Tower Height (ft)
0.9	2500	308/24
1.0	3500	495/30
2.3 (grid)	6212	687/50
3.0	7850	935/65
3.0 (battery charger)	8150	11,41/80

Note: Towers are supplied in kit form, without steel pipes. Installation costs for turbines and towers not included in retail cost estimates. Some suppliers charge for delivery and insurance if delivery sites are distant from suppliers' premises.

Table 1.4   2009 Total Installed Cost for 5 kW Wind Turbine

Component	Cost ($)
Turbine	5,900
Tower (40 ft)	4,600
Installation of turbine	400
Installation of tower	150
DC-to-AC inverter	2,000
Miscellaneous accessories	500
Total	13,550

1.6.6 Wind Turbine and Tower Installers

Wind turbine customers who do not like installation costs quoted by manufacturers can contact installation companies. However, the author recommends that users and wind turbine purchasers insist on having installation handled by the turbine and tower suppliers because they know their system components, installation priorities, and sequences. The following companies specialize in installation projects:

■ Solar Planet, Irvine, California, (949) 413-7888
■ Wind Monkey, Irvine, California, (949) 748-5774
■ Sustainable Power Systems, San Demas, California, (909) 394-4647

1.6.7 Potential Applications of Small Wind Turbines

Small-scale wind turbines are best suited for applications where electrical power consumption does not exceed 10 kW, for example:

- Remotely located homes
- Telecommunication transmitter sites
- Offshore platforms
- Cathodic protection
- Performance monitoring sites
- Water pumping
- Utility-connected homes and businesses
- Remote military posts.

1.6.8 Intermediate-Scale Wind Turbines

The output capacities of these turbines range from 20 to 100 kW and they are best suited for applications where small-scale and utility-scale turbines do not provide the most cost-effective operations in the long run. Intermediate turbines are widely used for on-grid and off-grid applications such as distributed generation, telecommunications, village electrification, and water pumping. Several U.S. companies manufacture wind turbines with moderate capacities. Energy Maintenance Service produces 35-kW and 65-kW turbines.

Another intermediate turbine supplier is the Entegrity Wind System which manufactures a 50-kW (Model EW-15) best suited for hybrid applications where the end user directly consumes wind-generated power. The EW-15 is best suited for farms, schools, industries, and municipal wastewater treatment facilities in grid-tied applications. For remote communities, the EW-15 unit can be incorporated into a utility system to provide electricity to small villages and towns at minimum cost. This system is capable of delivering 150,000 to 200,000 kilowatt hours per year, depending on the wind regime at the turbine site. The expected turbine life is about 30 years.

The 20-kW wind turbine manufactured by Wind Turbine Industries Corporation is capable of providing electric power for a broad range of applications such as Grid Intertied or hybrid battery charging systems at remote locations. The rotor size is about 8.8 meters or 29 feet and the system offers quiet, clean, reliable, and efficient operation.

1.7 Wind Farm Developers

Wind farm companies play critical roles in recommendations for and selections of appropriate installation sites for cost-effective operation of wind turbines needed to generate large amounts of electrical power (about 500 MW). A wind farm will contain several 2.5-MW turbines operating in parallel to generate large amounts of power, which will be more cost-effective to operate than a coal-fired plant with an identical power rating. Figure 14 shows a typical wind farm configuration of ten 2.5-MW turbines.

1.7.1 Wind Farm Dealers

Some companies and dealers specialize in site selection for installation of wind turbines to meet specific electric power generating needs on a specific area of land. Such dealers employ expert site survey experts to ensure proper installation site selection. These experts examine the terrain features, wind speeds and directions, and other relevant parameters that may impact the cost-effective operation of a wind farm. Such companies operating in the U.S. include Wing Energy Consulting and Contracting, Catamount Energy, Blue Water Wind, Action Energy, PPM Energy, Blue Sky Wind, Endless Energy Corporation, Cape Wind, Midwest Renewable Energy Corporation, Horizon Wind Energy, and Clipper Wind.

1.7.2 Renewable Energy Professionals

It will be of great interest to prospective customers to seek professional help or advice prior to selecting a Solar Panel Installation or Wind Turbine dealer to meet his electricity requirements with minimum cost and complexity. Green Professionals Directory provides the names, addresses and contact phone numbers of such professionals, which can be contacted if needed. Renewable energy professionals include:

- FSL Energy Management, Black Mountain, North Carolina, (828) 669-5070
- Southern Energy Management, Raleigh, North Carolina, (919) 836-0330
- Blue Ridge Energy Solutions, Morganton, North Carolina, (800) 869-8824
- Appalachian Energy Solutions, Sugar Grove, North Carolina, (828) 773-9762
- Solar Dynamics, Asheville, North Carolina, (828) 665-8507
- Sundance Power Systems, Mars Hill, North Carolina, (828) 689-2080

1.8 Design Configurations

Various design configurations of small wind turbines have potential applications where uninterrupted supplies of electricity with high load factors are required regardless of wind conditions. A hybrid version of a residential off-wind turbine offers the most cost-effective operation and reliability. Figure 1.5 shows a residential wind turbine configuration with off-grid capability, a home wind turbine configuration with

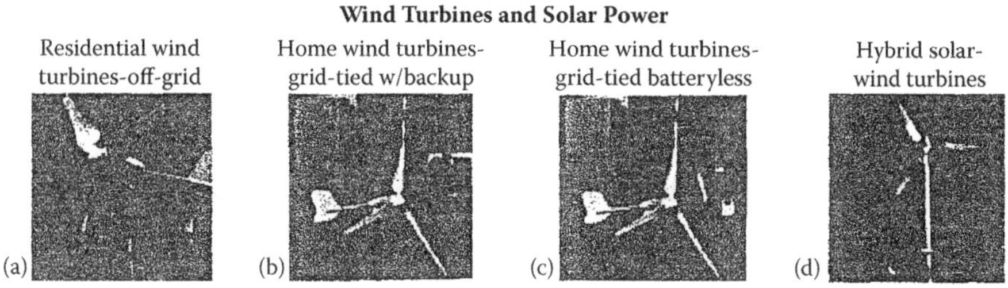

Wind Turbines and Solar Power

Residential wind turbines-off-grid	Home wind turbines-grid-tied w/backup	Home wind turbines-grid-tied batteryless	Hybrid solar-wind turbines

Figure 1.5   Residential wind turbine configurations.

grid-tied and back-up capability, a home wind turbine configuration with grid-tied and battery-less operation, and hybrid solar–wind turbine design configuration.

1.8.1 Residential Design Configurations with Off-Grid Capabilities

This wind turbine configuration is shown in Figure 1.5(a). This design exhibits improved overall performance, high efficiency, enhanced reliability, and superior workmanship. It works impressively in both high and low wind speed conditions. Its design includes a patented twisted high-efficiency blade, extremely efficient aerodynamics, optimum angle of attack along the entire blade length at variable wind velocities, highest lift-to-drag ratio, electromagnetic braking, over-speed control mechanism, and aerodynamic blade speed limitation. These turbines are rated below 1 kW. They are sold in kit form and are easily assembled and installed. The turbines are economical and relatively reliable.

1.8.2 Residential Design Configurations with Grid-Tied and Back-Up Capabilities

This configuration is depicted in Figure 1.5(b). Compared to batteryless grid-tied wind turbines, grid-tied devices with battery back-ups provide uninterrupted electricity at minimum cost. They also offer high reliability, energy independence, and back-up power for remote off-grid sites and under unstable grid conditions. If one has both wind and solar power systems, only one inverter is needed for a grid-tied wind turbine with battery back-up.

1.8.3 Residential Design Configurations with Grid-Tied and No-Battery Operation

This configuration is shown in Figure 1.5(c). Grid-tied wind turbines offer design simplicity, slightly better system performance, and easier set procedure than grid-tied systems with battery back-up capability. However, they lack back-up capability; when the grid goes down, these systems also go down. This is the most serious drawback of this design.

1.8.4 Hybrid Solar-Wind Turbine System Configurations

The hybrid solar–wind power system configuration is shown in Figure 1.5(d). This configuration generates electricity simultaneously from both solar and wind energy at operating conditions. It provides high efficiency, sustainability, and reliability. The system is light-weight, can be installed easily, and has an electromagnetic over-speed control mechanism that is more reliable than the traditional furling system. Only one inverter unit is needed for both solar and wind power modules, thereby making the hybrid cost-effective, clean, and reliable. Critical components of a small wind–solar hybrid system configuration are shown in Figure 1.6.

This system can be used for a variety of applications including utility-connected homes and businesses, remote homes, water pumping, telecommunications, and offshore platform lighting. Small wind turbine system suppliers claim that this hybrid is best suited for battery-charging applications in all climatic environments. According to published reports, more than 50 countries manufacture an estimated 50,000 small wind turbine generators for various applications.

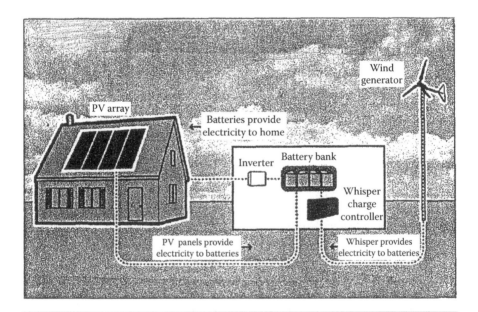

Figure 1.6 Critical components of small wind–solar hybrid system.

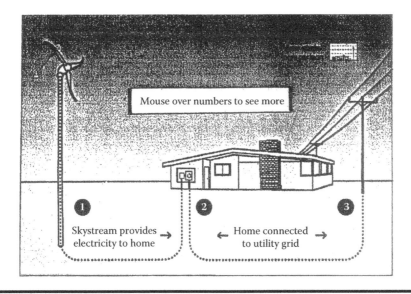

Figure 1.7 Skystream residential system that works in tandem with an electric utility supplier.

1.8.5 Compact Wind Turbine and Energy Systems for Dual Applications

Southwest Wind Power supplies a compact and inexpensive small-scale residential wind turbine called the "Skystream" that works in tandem with an electric utility supplier. This compact configuration shown in Figure 1.7 provides electrical energy to both residence and utility grid. In brief, when the wind is not blowing, the turbine provides quiet and clean electricity. When wind speeds are high, the system generates more electrical energy than the residence needs and the excess electricity is fed back to the grid. The electric meter will actually spin backward; the residential user in essence sells electricity back to the utility company. A compact wind turbine energy system as shown in Figure 1.7 will generate electricity for a residence and utility grid and be eligible for tax rebates and incentives provided by state and federal governments.

1.8.6 Critical Electrical Components

In general, a wind turbine comes in kit form, complete with rotor blades, generator, inverter, wind generator magnets, charge controllers, and essential accessories. The tower represents a large component of the system and is delivered directly to the customer site. However, the user must maintain stocks to replace damaged or non-operating parts as required to maintain normal system operation. Most of the critical components used by residential off-grid wind turbines, home wind turbines with grid-tied and back-up capabilities, and home wind turbines with grid-tied and battery-free capabilities are available from Windex Greenpower4less and wind power supply stores. Components include wind generators; generator magnets made of rare earth neodymium, ceramic, samarium–cobalt, flexible, and aluminum–nickel–cobalt; wind generator blades; towers; solar panels for hybrid systems; inverters; home turbine kits; charge controllers; wind–solar water pumps; and 1-kW, 48-V small wind turbines.

1.9 Next Generation Wind Turbines with Unique Features

This section describes the next generation of wind turbines and their unique design features. The kinetic energy of the wind can be converted into electrical energy using highly efficient vertical blade designs.

1.9.1 Helix Wind Turbines

Helix Wind offers an elegant and cost-effective solution for homes and small businesses. The Helix turbines are small capacity devices. The next generation of small wind turbines will be powerful enough to meet electricity requirements yet harmonious with the environment. These small turbines allow the kinetic energy of the wind to be captured by the unique and highly efficient vertical blade design configuration.

1.9.2 Wind Turbines Operating from Ocean Surfaces

Scientific data from NASA's QuickSCAT satellite reveal that we can harness ocean wind energy to generate electricity. The maps generated by the NASA satellite can aid in planning and locating offshore wind farms for producing electrical energy. High-speed ocean winds are considered ideal for generating large amounts of electricity at lower cost. Ground roughness exerts critical effects on the vertical distribution of wind as shown by the curves presented in Figure 1.8. These effects are dependent on terrain features. The vertical distribution of the wind speeds experienced by a wind turbine operating from the ocean surface is based on turbine height. The vertical distribution of the wind follows an exponential law as a function of height as illustrated in the figure.

1.9.3 Wind Turbine Design Based on Jet Engine

A Massachusetts aerospace company has devised a new concept that may be at least twice as efficient as traditional turbines deploying rotor blades that force air around

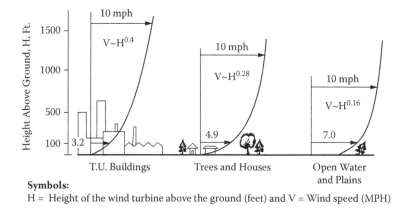

Symbols:
H = Height of the wind turbine above the ground (feet) and V = Wind speed (MPH)

Figure 1.8 Vertical distribution of wind dependent on ground objects and terrain features. The vertical distribution of the wind experienced by a turbine operating from an ocean surface depends on turbine height.

them instead of through them. The device channels wind into a vortex that spins the blades and generates electricity. The Massachusetts's company hopes to have a prototype turbine completed by the end of 2009. Unfortunately its website (flodesignwindturbine.org) is not currently functional.

1.9.4 Vertical Axis Wind Turbines

Tom J. Gilmour, a Canadian inventor, recently published a new design for a wind turbine; a patent application for the design has been filed. The inventor hopes to reserve worldwide patent rights for his design and believes that his windmill will be most complex ever designed. He is optimistic about the feasibility of his design and plans to build a working model in near future. His website (Tom's Whirligig) provides detailed descriptions illustrations, Google Sketchup files, and a system block diagram.

1.9.5 Floating Offshore Wind Turbines

StatoilHydro, a Norwegian company, is working on a 2.3-MW floating wind turbine that will be attached to the top of a "Spar-buoy" and moored to the sea bed by three anchor points to provide stability to the system. The company planners predict that the 80-m diameter, 65-m high Siemens turbine will have lower deepwater installation costs compared to a traditional turbine installation on a fixed platform. The company plans to have a working prototype completed by the end of 2009 and thereafter feasibility tests will be performed to determine reliability and performance in high wind environments.

Table 1.5 Wind Power Classification Parameters

Wind power density at 50 m height (W/m²)	Wind Speed (m/sec)	Potential Status
300 to 400	6.4 to 7.0	Fair
400 to 500	7.0 to 7.5	Good
500 to 600	7.5 to 8.0	Excellent
600 to 800	8.0 to 8.8	Outstanding
800 to 1,600	8.8 to 11.1	Optimum

1.10 Typical Wind Power Estimates for United States

According to the reports from National Renewable Energy Laboratory, site selection for the installation of wind turbines depends on wind power calculated as a function of turbine height from the ground. Preliminary data must be screened to eliminate areas that will not generate electrical energy at reasonable costs. Furthermore, such data will eliminate potential installation sites unlikely to be developed onshore due to specific land uses or environmental issues. Wind power classification is of paramount importance prior to selecting a site for turbine installation. Based on an operating height of 50 m above the ground, estimated values of wind power density in watts per square meter of surface area (W/m²) for various wind speed zones are summarized in Table 1.5. Note the wind speeds shown are based on a Weibull parameter *k* value of 2.0 that is widely used in preparation of U.S. wind resource maps [3].

It is not possible to estimate the wind power classification data for various locations around the world. Each country must generate such data to determine optimum wind turbine location sites within its territory. At present such data is not available for various countries.

1.11 Summary

This chapter describes the history of wind turbine operations around the world with emphasis on installation dates, performance parameters of various types of wind turbines, design aspects of critical components, and installation requirements for turbines and towers. Major benefits and disadvantages of alternate energy sources are summarized. Site selection requirements are discussed. Repowering schemes to yield more electrical energy from existing wind turbines are briefly described. Countries that participated in the early development of wind turbines are identified. Critical system components and their important functions are discussed, with emphasis on reliability and longevity.

Most popular wind turbines such as utility-scale and small-scale types are described along with details about efficiency, reliability, and tower requirements. Tower height as a function of wind speed to capture maximum kinetic energy at various locations is discussed. Pay-off periods for utility- and small-scale scale wind turbines are specified and various design configurations for small-scale wind turbines are identified. Hybrid solar–wind energy systems are briefly described. Installation cost break-down for a 5-kW small-scale design is detailed. Locations and phone numbers for various wind turbine suppliers and manufacturers and wind farm development companies are provided for individuals interested in procuring wind turbines. The next generation of wind turbine designs is summarized related to performance improvement, noise reduction, and installation cost reduction. Wind power classification status in terms of electrical energy generated per unit surface area occupied by the operating system (watts per square meter) is provided for a turbine height of 50 m or 164 ft.

References

[1] Europe replaces old wind turbine farms, *IEEE Spectrum*, January 2009, p. 13.
[2] Ground zero for alternative energy, *Los Angeles Times*, March 1, 2009, p. A-20.
[3] Wind power classification, *Microwave Magazine*, February 2009, p. 35.

Chapter 2

Design Aspects and Performance Requirements

2.1 Introduction

This chapter will describe various types of wind turbines with particular emphasis on design requirements and performance capabilities. Extraction of wind energy from wind turbines depends strictly on wind speed and direction. Despite the limited availability of suitable operating sites with high annual wind energy potential, deployment of wind turbines to generate electricity merits serious consideration to reduce dependence on fossil oils. Note that the kinetic energy produced by a wind turbine is in a thermodynamic sense the most desirable type of energy because it is readily available and can be converted in any other form of useful energy.

First we should identify potential applications of the wind energy captured by turbines. A wind turbine drives an electrical generator whose output can be stored in batteries or linked to a utility power grid. Wind turbines can be used to produce clean fuel composed of hydrogen and oxygen, develop fuel cells, drive heat pumps, or operate air compressors and water pumps. In brief, the energy generated by wind turbines can be converted into electrical, chemical, thermal or potential energy. The most viable application is pumping water to reservoirs associated with hydroelectric power plants. For any application, it is vital to investigate whether a specific wind turbine offers high efficiency, year-round reliable operation, and lower operating costs.

It is important to note that linking with a utility power grid requires synchronization with the grid frequency. That means a wind turbine must be operated at a constant angular rate or must drive a variable input speed constant frequency output electrical generator. Such machines are available with power output capabilities ranging up to 60 kW but are prohibitively expensive. Storage of the electrical energy generated by the wind turbines is also expensive because of the cost of batteries and the additional equipment needed to convert the alternating current to direct current, compatible with battery charging requirements. Furthermore, the efficiency of the storage and retrieval of energy is about 75%, which means that storage of wind turbine-generated electrical energy may not be cost-effective.

2.2 Types of Wind Turbines

Several types of wind turbines operate in various regions of the world. Figure 2.1 illustrates early and modern wind turbines. The windmill in the figure has played a key role in energy generation, particularly in water pumping applications. It is interesting to note that more than one million wind wheels continue to pump water in different parts of the world. Wind turbines are classified into two major categories: horizontal axis wind turbines (HAWTs) and vertical axis wind turbines (VAWTs) as illustrated in Figure 2.2 and Figure 2.3, respectively.

The HAWT category includes wind turbines with upwind rotors, wind turbines with downwind rotors, windmill turbines, wind wheel turbines, and turbines with high tip speed ratios ranging from 1 to 8. The three distinct types of VAWTs are those with Savonius rotor configurations, Darrieus wind turbines, and Giromill wind turbines.

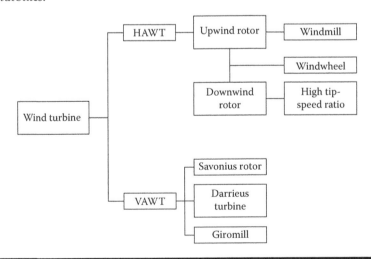

Figure 2.1   Windmill for water pumping applications.

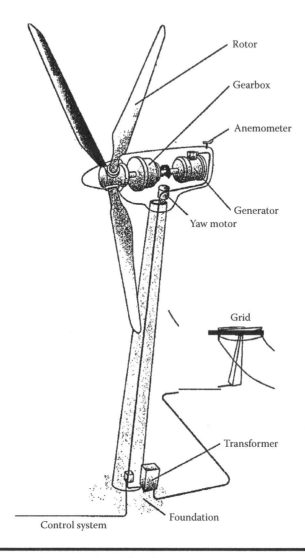

Rotor

Gearbox

Anemometer

Generator

Yaw motor

Grid

Transformer

Control system

Foundation

Figure 2.2   Horizontal-axis wind turbine (HAWT).

Performance capabilities and limitations of these turbines, along with details of system and installation costs, reliability, and safety will be summarized below. Wind turbines using specific rotor concepts are shown in Figure 2.4. The figure also displays power coefficients as functions of blade tip speed-to-wind speed ratio for propeller type windmills, turbines using high-speed two-bladed rotors, turbines with multi-bladed rotors, turbines with Savonius rotor configurations, and Dutch type turbines.

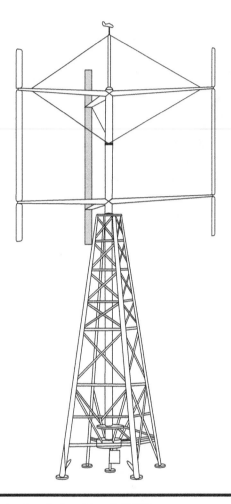

Figure 2.3   Vertical-axis wind turbine (VAWT).

2.2.1 Windmills

The windmill features a very robust design composed of simple and inexpensive components. The system is easy to operate and requires minimum repair and maintenance. The greatest advantage of a wind wheel is shown in Figure 2.1. In comparison to a conventional wind turbine with a few slender blades rotating at high speeds, a windmill starts more easily because the rotor blades cover a far larger share of the swept area. The windmill is an effective device for pumping water because a lot of water is required to get it running.

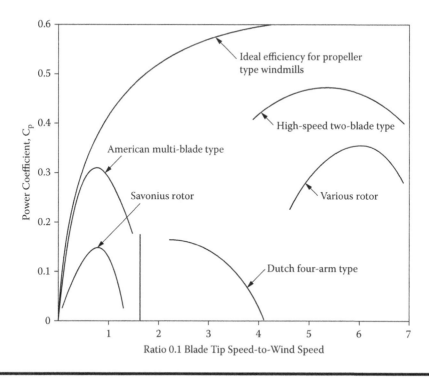

Figure 2.4   Rotor concepts and power coefficients.

2.2.2 Farm Type and Dutch Type Wind Turbines

The farm type and Dutch type wind turbine configurations are depicted in Figure 2.5. The two turbines have excellent starting torques and power coefficients ranging from 0.15 to 0.25. Modern turbines using propellers exhibit poor starting torques and power coefficients between 0.40 and 0.50 and deploy propellers with two- or three-blade configurations as illustrated in Figure 2.4. As mentioned earlier, modern wind turbines are classified into horizontal axis wind turbine (HAWT) and vertical axis wind turbine (VAWT) categories. Each category is further subdivided based on upwind rotor configuration, downwind rotor configuration, high tip speed ratio, Savonius rotor design, windmill design, and wind wheel design.

2.3 Modern Wind Turbines

Modern types of wind turbines including the HAWT and VAWT machines will be discussed with emphasis on applications and performance capabilities. A modern wind turbine, regardless of the type, consists of basic components: a base or

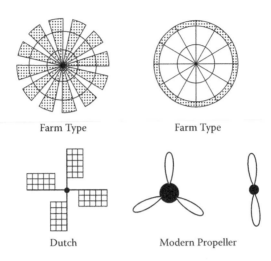

Farm Type   Farm Type

Dutch   Modern Propeller

Figure 2.5   Farm type and Dutch type wind turbines.

foundation, tower, nacelle enclosure housing a generator, gearbox, yaw motor and other electrical mechanical accessories, rotor, control system, and transformer. The turbines that dominate the current markets operate at high tip speed-to-wind speed ratios ranging typically from 5 to 7 and utilize three-bladed rotors with rotational speeds of 10 to 30 revolutions per minute (RPM). Most manufacturers offer several models with different hub heights and rotor diameters to tailor installation to meet customer site and power output requirements.

2.3.1 Horizontal Axis Wind Turbine (HAWT)

If the rotor blades are connected a horizontal shaft, the device is an HAWT. Such turbines are widely used for commercial applications. Critical components are depicted in Figure 2.2. The gear box and generator are connected to the horizontal shaft and the transformer is located at the base of the tower as shown in Figure 2.2. A horizontal axis wind turbine may be of rotor-upwind upwind design to face the wind or rotor-downwind design to enable the wind to pass the tower and nacelle before it hits the rotor. Most modern wind turbines have upwind design configurations and range from prototypes in the MW class to smaller turbines with nominal power output capacities of 20 to 150 kW. Nineteenth century water pumping wind wheels deployed downwind rotors. Major design efforts are directed toward the major components such as rotor diameter, number and twist angle of rotor blades, tower height, rated electrical power, and control strategy.

The tower height for HAWTs is extremely important because wind speed increases with the height above the ground. Rotor diameter (D) is equally important because it determines the area (A) needed to meet specific output power level.

As mentioned, HAWT systems are best suited for electrical power generation and micro-turbines composed of two to six rotor blades are most attractive for battery charging applications. Since the 1980s, several versions utilize grid-connected wind turbines consisting of two or three rotor blades, some with the rotor-downwind and others with the rotor-upwind configurations. The major advantage of rotor-downwind design is its automatic adjustment to wind direction—an important safety and operational feature. However, field operating data indicate that this adjustment is not possible under abrupt or sudden changes in wind direction. This operational deficiency can be overcome with a three-bladed upwind rotor design configuration. For this reason upwind rotor designs dominate the current markets.

The power output performance of a HAWT can be optimized by selecting a ratio between the rotor diameter (D) and the hub height (H) very close to unity. The rated power output of a wind turbine is the maximum power allowed for the installed electrical generator. The control system must ensure that this power is not exceeded in high-wind environments to avoid structural damage to the system. HAWT systems typically deploy two or three rotor blades. A turbine with two rotor blades is cheaper, but it rotates faster, thereby producing a visual flickering effect; also the aerodynamic efficiency of a two-blade rotor is lower than that of a three-blade rotor.

Two-blade wind turbines are often downwind installations. In contrast, a three-blade HAWT offers smooth operation and is therefore less disturbing. Two-bladed wind turbines are normally downwind machines, i.e., the rotor is downwind on the tower. Furthermore, the connection to the shaft is flexible because the rotor can be mounted through a hinge. This is called a teeter mechanism; it prevents transmission of the bending moments from the rotor to the mechanical shaft. This type of construction is more flexible than a stiff three-bladed rotor configuration and some of its components are more compact and lighter, thereby reducing the procurement cost. Ensuring the stability of the flexible rotor is essential. Downwind turbines are noisier than upstream turbines; the once-per-revolution tower passage of each blade is heard as a low-frequency noise due to human ear sensitivity in the low-frequency region.

2.3.2 Vertical Axis Wind Turbine (VAWT)

The critical components of a VAWT are not shown in Figure 2.3 because of the configuration complexity. The greatest advantage of a VAWT is that the generator and gear box can be installed at the base of the tower, thereby making these components easy to service and repair. Both the Savonius and Darrieus turbines fall into this category and are available commercially. However, these turbines have small output capacities and hence are widely used for low-power applications such as battery charging in areas where power grids are not available.

George Savonius, a Finnish designer and inventor, developed the first VAWT in 1924 and it utilized what is now known as the Savonius rotor design configuration. This turbine consists of a vertical S-shaped surface that rotates around a central axis. VAWTs are not as efficient as HAWTs. Savonius rotors are most commonly seen as advertisement devices erected in front of businesses to increase sales. Savonius rotors have very limited power outputs and poor efficiencies, but they are reliable and easy to maintain. Savonius turbines with high capacities require large amounts of materials and are not cost-effective over the long run.

Power coefficients for various turbine rotors are illustrated in Figure 2.4. It is evident from this figure that a Savonius has the lowest power coefficient and it operates only over a specified blade tip speed-to-wind speed ratio. Note that the Darrieus rotor has a power coefficient close to 0.35 at a blade tip speed-to-wind speed ratio ranging from 5.5 to 6.5. Because of their limited capacities and lower efficiencies, these turbines are best suited for battery charging applications for telecommunication masts and remote lighthouses. The Savonius turbine is widely used as a starter motor for a Darrieus turbine because the Darrieus lacks self-starting capability under certain wind speed conditions.

2.3.3 VAWT Operating Requirements

A VAWT requires certain wind direction conditions. VAWTs overcome the difficulties of propeller-based windmills that must be pointed into the wind by orienting the axis of the rotation so that the full force of the wind can be sensed from any direction. It is important to mention that a VAWT employs an aerodynamic lifting surface to generate torque (T). Darrieus and Giromill turbines belong to this category. Such turbines may have skipping rope-curved blades or blades with variable pitch. Variants of the crosswind axis types have been proposed, but their superiority over the current turbine designs has not yet been demonstrated.

2.3.4 Advantages and Disadvantages of VAWTs

Major advantages of a VAWT are:

- Elimination of yawing mechanisms
- High efficiency
- High rotational speeds that offer reduction in gear ratios

In the case of Giromills with articulated blades, it is necessary to sense the direction of the wind in order to time the cyclic pitch changes accurately. The disadvantages of a VAWT are:

- Tendency to stall under gusty wind conditions
- Low starting torque
- Dynamic stability problems
- Sensibility to off-design conditions
- Low installation height limiting operation to lower wind speed environments

2.3.5 Operational Difficulties of VAWTs

Rotation of the blades around the axis changes both the angle of attack of the aerodynamic surface and the local dynamic pressure of a VAWT. In addition, the torque produced by the rotor is not steady. Thus, the load driven by the wind turbine experiences a fluctuating power input that generates a fluctuating turbine output. Furthermore, in case of Darrieus rotors and straight-bladed turbines without blade articulation mechanisms, regulation of angular rate must be performed by the load.

Dynamic analysis of a VAWT is extremely difficult because of the variability of local air flow conditions as the blades rotate around the axis. A VAWT produces both drag and lift in the direction normal to the wind vector. Under these circumstances, the resulting flow field is very complex and it is very difficult to make accurate estimates of the inflow conditions. In the windward portion of the rotation cycle, the blades move into a skewed flow field that undergoes very rapid variation within the installation site. In one quadrant of the windward portion of the rotation cycle, the blades advance into the wind and retreat in the other quadrant. In the leeward region of the rotating cycle, the blades operate in the flow field influenced by the blade motion in the windward phase of the operation. Under these circumstances, the extremely complex flow situation is beyond the existing analytical capabilities.

2.3.6 Simplified Procedure for Predicting Darrieus Wind Turbine Performance

Using the propeller theory and equating the blade forces with wake common momentum defect, axial interference in flow velocity and calculated power coefficients can be calculated using first-order analytical techniques. The calculation will yield approximate results because the analysis is based on a large number of assumptions which, in some cases, contradict theoretical principles of fluid mechanics. This approach is as good as any other theories, but its ability to predict the performance parameters of a Darrieus wind turbine is not entirely satisfactory. A model based on this theory can be used to predict the power coefficient for a Darrieus rotor as a function of tip speed (x) and various values of initial drag coefficients. The tip speed ratio is the ratio of the product of blade radius and angular speed of the rotor to the wind velocity. Computed values of power coefficient for a Darrieus rotor are summarized in Table 2.1.

2.3.7 Understanding Flow Phenomena of VAWTs

Appreciation of the complexities involved in the flow phenomena associated with the operation of a VAWT can be obtained from a local flow vector diagram for a blade section. The flow relationships of the various blade parameters can be realized

Table 2.1 Calculated Power Coefficient Values of Darrieus Rotors

	Initial Drag Coefficient		
Tip Speed Ratio	*0.010*	*0.015*	*0.020*
3	0.370	0.375	0.382
4	0.433	0.402	0.374
5	0.435	0.378	0.353
6	0.400	0.305	0.221
7	0.315	0.162	0.021
8	0.175	0.000	0.000

from a flow vector diagram for a blade section. Various parameters associated with a blade section include drag and lift coefficients, rotation wind vector, chord line, pitch angle, twist angle, local streamline velocity, and actual velocity components as a function of polar angle and twist angle. If a reasonable mathematical model could be developed for the distribution of the longitudinal and transverse interference factors, stream surface analyses connecting the forces on the blades with defects in wake momentum would probably yield reasonable approximations for power coefficients. This model will be useful for the analysis of constant local streamline velocity.

2.3.8 Early European Wind Turbines

In the early 19th century, small wind turbines with capacities far below 5 kW were developed by advanced European countries including Denmark, Germany, Belgium, and others. The early versions of wind turbines illustrated in Figure 2.5 had limited applications because of low power output capabilities and poor efficiencies. Farm types, Dutch types, and two-bladed turbines constitute the early versions.

2.4 Off-Design Performance

Off-design performance specifications offer cost-effective design configurations; off-design parameters are selected to provide high operating efficiencies at a wide range of wind speeds. This is desirable because wind speeds vary from location to location and are strictly dependent on climatic environmental parameters. Cost-effective operation of a propeller-based wind turbine at its design point means that system parameters must allow a minimum drag-to-lift ratio to be realized over a wide range of wind speeds.

2.4.1 Critical Design Aspects

The off-design concept offers optimum performance over a wide range of wind speeds, even when the angular rate of the turbine may vary if the demands of electrical loads are compatible. However, under these operating conditions, the variations in Reynolds numbers introduce deviations from the anticipated optimum turbine performance, but these are of secondary importance in turbines with high capacities. When wind turbines are used to generate large amounts of electricity, it is necessary to deliver the electrical power at a fixed voltage and frequency compatible with utility grid operating requirements. Large wind turbines are designed to link to main electrical power grids and thus the frequency must be synchronized precisely.

Large propeller-based wind turbines achieve such control by implementing blades with pitch variability. Small wind turbines with fixed angles operate at off-design conditions whenever wind speeds deviate from the nominal value. This situation may be alleviated somewhat by the recently developed, electronically controlled, variable input speed, constant frequency output generators. Because of high costs, deployment of such generators is limited to medium wind turbines with outputs from 50 to 60 kW. These generators are widely used in military and commercial aircraft.

2.4.2 Impacts of Variations in Design and Operating Parameters

Wind turbine performance parameters are affected by variations in tip speed ratio, fixed pitch angle, and angular rate. Preliminary studies performed by the author indicate that variations of the power coefficients of fixed pitch wind turbines occur with variations in tip speed ratios. The studies further indicate that a tip speed ratio greater than unity means that a constant angular rate turbine is operating at wind speeds below the nominal value. It is interesting to note that considerable latitude exists in the combinations of airfoil characteristics and tip speed ratios that will result in high operating efficiencies over a wide range of wind speeds.

2.4.3 Impact of Drag and Lift Coefficients on Maximum Power Coefficient

Maximum power coefficient for a turbine rotor is the most critical performance parameter [2]. This coefficient determines the optimum performance of a turbine under all operating and design parameters. Maximum power coefficient for a turbine rotor can be achieved at increasing values of the tip speed ratios as the drag coefficient-to-lift coefficient ratio (C_D/C_L) decreases. These maxima tend to be less sharply defined at lower values of this ratio, but are readily discernible. Designs of wind turbines incorporating some selected airfoil sections with a particular drag-to-lift coefficient ratio over the entire length of a blade would select tip speed ratios corresponding to the maximum values of system performance.

The operation of such turbines under off-design conditions presents an interest engineering problem that heavily influences the overall system design considerations. For airfoil sections capable of achieving minimum drag-to-lift ratios, the angle of attack must be selected close to 0.1 radian or 5.73 degrees. The three distinct values of this ratio of 0.01, 0.03, and 0.05 are associated with design tip speed ratios of 5.0, 3.5, and 3.0, respectively, that correspond approximately to the maxima.

Furthermore, a variation of the power coefficient of a turbine may occur through variation in the tip speed ratio; this illustrates the extreme sensitivity of relatively inefficient wind turbines to variations in operating conditions. Curves representing the power coefficient versus the tip speed ratio depend strictly on the airfoil sections and demonstrate the trend of the power coefficient. These curves will indicate the off-design procedures that must be followed to arrive at an acceptable design for achieving optimum performance.

2.4.4 Performance Enhancement Schemes

It is important to remember that the kinetic energy of the wind is used to drive a wind turbine. Both the low energy density of the winds and the relatively low theoretical upper limit on the extraction of wind power have generated considerable interest in augmentation of the power output levels of the rotors. Turbine designers have proposed four distinct augmentation schemes:

- Concentrator
- Aerodynamic lift auxiliary
- Ejector and diffuser
- Shrouds

Each augmentation scheme includes an almost unlimited number of variations as new concepts appear continuously. Typical configurations channel large quantities of flow into the rotor assembly and thus increase its output power because both flow rate and flow speed are increased. At first glance, this appears an easy method for increasing the output power of a turbine rotor. However, comprehensive examination revealed that the overall momentum and the energy balances limit the power coefficient of the projected inlet area to the Betz-Lanchester limit of 16/27 or 0.5926 [2].

Because of the inlet and fractional losses, this limit cannot be achieved in theory because the air flow exiting the rotor must be diffused to ambient pressure, which sets the upper limit on the flow rates that can be handled. Concentrators must be viewed as techniques to gain additional power from a given rotor through the use of passive and inexpensive aerodynamic surfaces. However, the concentrator schemes involve additional weight that will add to tower and foundation cost and may force a turbine installer to consider another option for the ground plane location.

Certain augmentation concepts must be explored to improve starting torques and power coefficients over a wide range of tip speed ratios:

- Stationary surfaces: flow is channeled into the active elements; the main effect is to increase the wind energy capture area.
- Inlet ducts: dimensions of the inlet ducts must be selected to capture a larger stream tube or wind energy.
- Ejector ducts: duct dimensions must be selected to facilitate the mixing of spent flow from the turbine with the incoming wind stream and send the mixture through a diffuser in the anticipation that the average energy of the flow will be sufficient for diffusion to the free stream pressure.
- Diffuser ducts: they are attached downstream of the turbine in the hope that the spent flow will diffuse to ambient wind pressure more efficiently than it would on its own accord.

All these concepts are intended to provide more efficient and reliable operation of a wind turbine. However, on the basis of the projected area of the recommended device it is important to note that the upper limit on wind energy extraction is limited by the laws of physics. If the above devices are considered stationary but include orientation provisions to achieve a certain turbine power output with a small rotating element, then their implementation may be justified. For example, a 10-ft radius wind turbine with a 50-ft radius blade and a 285-ft long diffuser with a divergence angle of 8 degrees will not likely be competitive with a 50-ft radius simple propeller. The precise alignment of such a device to ensure optimum wind input poses a serious and risky challenge. In addition, visual impacts will eliminate various forms of shrouds from serious consideration. While some of these augmentation devices offer some improvement in turbine performance, the additional cost and weight imposed on tower and foundation represent serious disadvantages.

2.5 Techniques for Capturing Large Amounts of Wind Energy

Techniques capable of capturing the large amounts of wind energy will be investigated with emphasis on costs and safety factors. Selection of an ideal location for a wind turbine installation is very important. The selection process must consider factors such as wind speed and direction, desirable terrain features, nearby residential areas, and annual energy capture. Note that the ratio for actual energy captured by a wind turbine to that which could be captured is very critical. Furthermore, the wind speeds must be within the optimum range throughout the year at the designated location to enable the turbine to operate at its maximum power coefficient. To meet this operational criterion,

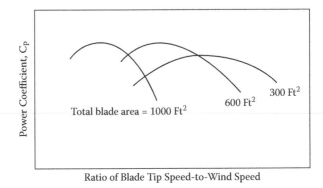

Figure 2.6 **Power coefficient as function of total blade area and blade tip speed-to-wind speed ratio.**

wind speeds from 20 to 30 m/min are recommended by installers—this is the most important site selection requirement. Other selection parameters include installation height, blade parameters, airfoil characteristics, and aerodynamic requirements; they all play important roles in efficient capture of wind energy by a wind turbine.

Power coefficient is dependent on several factors such as installation site features, rotor blade areas, angle of attack, flow rate, pressure drop at turbine, and other issues. Impact on power coefficient and power delivered due to rotor blade area, flow rate, and pressure drop at the turbine can be seen in Figure 2.7. Note that large blade areas yield both greater power outputs and improved power coefficients but over a narrow range of tip speed ratios as illustrated in Figure 2.6. Turbine blades with smaller areas provide lower power coefficients over a wider range of tip speed ratios.

Power Coefficient, C_p = Power Delivered/$\frac{1}{2}\rho A_T V_0^3$				
Turbine Blade Area	*Flow Rate*	*Pressue Drop at Turbine*	*Power Delivered*	C_p
Small	High	Small	Low	<0.59
Large	Low	Large	Low	<0.59
Optimum	Optimum	Optimum	Optimum	0.59
Symbols: p = air density (Kg/meter³); A_T = Turbine blade area; V_0 = Wind speed.				

Figure 2.7 **Impacts on power coefficients.**

2.5.1 Impacts of Blade and Angle Parameters on Performance Capabilities

It is important to mention that the pitch angle is variable in an articulated blade Giromill wind turbine, but no variation is present in a Darrieus rotor. Furthermore, the variation of the relative flow about the blade section can be seen in the four quadrants of the flow cycle for an identical local streamline velocity. Actual velocities vary significantly as a function of polar angle, particularly in the leeward portion of a cycle. The drag coefficient C_d for a blade can be written:

$$C_d = [C_{D,0} + (C_L^2/\pi A_R)] \tag{2.1}$$

where $C_{D,0}$ is the initial value of the drag coefficient, C_L is the coefficient of lift, and A_R is the area of the rotor. The slope of the lift coefficient C_L is strictly dependent on the angle of attack and its typical value is about 0.67 for a Darrieus rotor. The aerodynamic performance of the rotor blades is a function of Reynolds numbers; performance deteriorates at low Reynolds numbers. Since the Reynolds numbers range from low to high based on wind environments, the current analytical techniques in vertical axis wind turbine engineering are far from satisfactory.

The important performance parameter for a VAWT is the power coefficient, which is strictly a function of the tip speed ratio (x). The expression for the tip speed ratio can be written as:

$$X = [Rw/V] \tag{2.2}$$

where R is the blade radius, w is the angular velocity of the rotor, and V is the wind velocity experienced by the rotor blades. The equation for the power coefficient [1] for a vertical axis wind turbine is very complex and can be written as:

$$C = (Nc/2R)\,(Xw)\,[C_L \sin(\theta) - (C_{D,0}+C_L^2/\pi A_R)\,(x + \cos(\theta)] \tag{2.3}$$

where N is the number of rotor blades, R is the blade radius, C is the local blade cord, X is the tip speed ratio and is equal to (Rw/V), x is equal to 1.5X, and θ is the polar angle that exerts a significant impact on the velocity variations in the leeward portion of the cycle. Various parameters involved in Equation (2.3) have typical values as follows:

$$C_L = 4.0,\ C_{D,O} = 0.01,\ \pi A_R = 30\ \text{and}\ Nc/2R = 0.075$$

Inserting these parameter values in Equation (2.3) allows calculation of power coefficients for a VAWT. Calculated values of power coefficient [2] for a VAWT at an optimum as a function of tip speed ratio are shown in Table 2.2.

Table 2.2 Calculated Values of Power Coefficients at Optimum Drag-to-Lift Ratios for VAWTs

Tip Speed Ratio (X)	Power Coefficient
2	0.247
3	0.347
4	0.418
5	0.435
6	0.416
7	0.335
8	0.168

It is important to note that the values of the power coefficients for VAWTs are generally lower than those for HAWTs for a given drag-to-lift ratio (ε) because of the reasons cited above. In brief, VAWTs are generally less efficient than HAWTs.

Power coefficients for HAWTs are higher than those for VAWTs irrespective of the drag-to-lift ratios and tip speed ratios. In a HAWT case, angle of attack (α) plays a critical role in shaping performance. An angle of attack equal to 0.1 radian or 5.73 degrees offers optimum performance for a HAWT. Calculated values of power coefficients for a HAWT as a function of tip speed ratio (X) and drag-to-lift ratio (ε) are shown in Table 2.3.

Note that the power coefficient for a horizontal axis wind turbine still maintains an acceptable value even when the tip speed ratio varies from 2 to 10. This is the most critical design issue for choosing between VAWT and HAWT systems. It is evident from tabulated data [2] that the maximum power coefficient of 0.45 occurs when the tip speed ratio has a value of 3 and the drag-to-lift ratio equals 0.05. Furthermore, at the 0.05 drag-to-lift coefficient (ε), acceptable power coefficient values are possible only over a very limited range of tip speed ratios. At a drag-to-lift ratio of 0.03, acceptable power coefficient can be achieved over a tip speed ratio between 2 and 5.

Note that at a drag-to-lift ratio of 0.01, the power coefficient retains a minimum value of 0.5 over a wide range of tip speed ratios. It is evident from these data that as power coefficient of 0.55 and better is possible when the drag-to-lift ratio is maintained at 0.01. All these calculated values of power coefficient assume an angle of attack equal to 0.1 radian or 5.73 degrees. Based on these findings, we can conclude that maximum values of power coefficients are possibly only for HAWT systems operating over a wide range of tip speed ratios when the angle of attack is kept close to 5.73 degrees

Table 2.3   Power Coefficient (C$_P$) as Function of Various Parameters of HAWTs

	Drag-to-Lift Ratio (ε)		
Tip Speed Ratio X	0.05	0.03	0.01
2	0.44	0.46	0.38
3	0.45	0.51	0.53
4	0.43	0.50	0.55
5	0.27	0.45	0.56
6	–	0.34	0.55
7	–	–	0.53
8	–	–	0.49
9	–	–	0.44
10	–	–	0.31

and the drag-to-lift ratio equals 0.01. In addition, maximum annual capture of wind energy is possible only when a wind turbine can operate over a wide range of tip speed ratios. One can now visualize the advantages of HAWT installations.

2.5.2 Techniques for Achieving High Power Coefficients

A selected installation site must provide a blade tip-speed-to-wind speed ratio between 5 and 7 throughout most of the year to catch maximum wind energy needed for higher power coefficients. Studies performed by the author indicate that the power coefficient is dependent on turbine blade area, flow rate, pressure drop at the turbine, wind speed, and wind energy captured. The studies further indicate that power coefficients depend strictly on blade area and blade tip speed-to-wind speed ratios. Power coefficients for wind turbines as a function of total blade area and blade tip speed-to-wind speed ratio are shown in Figure 2.6. The overall impacts on the power coefficient from various parameters can be seen in Figure 2.7.

2.5.3 Installation Site Requirements for Optimum Performance

Most desirable installation sites for wind turbines include hills, ridges, and high-velocity wind flow regions such the Gobi Desert in China. Impacts of terrain features on desirable locations for wind turbines are shown in Figure 2.8. Note the exit pressure coefficient and interference factor to some extent are dependent on the

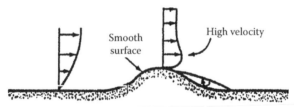

(a) Well rounded hill or ridge (suitable site)

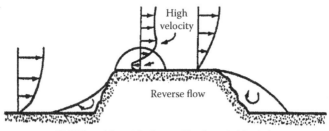

(b) Hill or ridge with abrupt sides (unsuitable site)

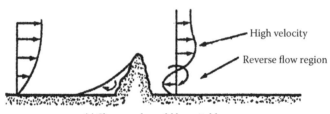

(c) Sharp peak-could be suitable

Figure 2.8 Impacts of terrain features.

turbine installation parameters. The power coefficient is also dependent on these two parameters as illustrated in Figure 2.9.

2.5.3.1 Wind Parameters over Ridges and Hills

Significant improvements in speed and direction may be noted as wind approaches hills and ridges. It is critical to mention considerable variations in wind speed as wind travels over ridge sections. Studies by wind turbine designers indicate that wind speed increases near the summit of a long ridge lying across the wind path. The winds accelerate as they pass over the ridge, and accelerate even more when they pass around the end regions of the ridge [3]. The studies further indicate that the wind speed increases as it approaches the front of the ridge and attains its highest speed as it reaches the peak.

Increases in wind speeds close 200% have been observed at the peak of a ridge; wind speeds may double as the flow accelerates up the gradual slope of a long ridge.

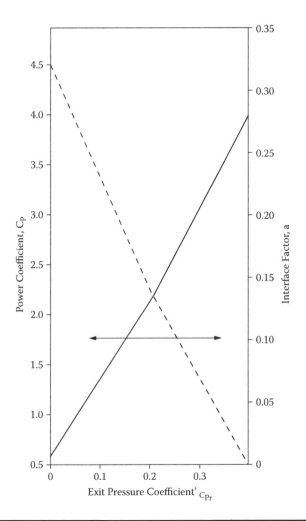

Figure 2.9 **Impacts of exit pressure coefficient and interference factor on power coefficient.**

Farmers living in ridge regions observed that this enhancement occurs only in the top third of the slope near the crest. Mountain passes can accelerate winds up to 100 mph (50 m/sec) during storms associated with the cold fronts. Regions with higher wind speeds are considered prime sites for wind turbine installations and are best suited for high-power installation. In brief, high wind speeds are highly desirable for wind turbines with output capacities ranging from 2.5 to 3.5 MW.

It is important to mention that average wind speeds of 20 to 30 mph are most desirable for safe operation of high-power turbines. Ridge tops may not only experience more frequent winds, but also have stronger wind environments. Terrain features that enhance wind speeds can also create turbulence that may wreak havoc on wind turbines. Wind turbines may succumb in consistently turbulent environments. Turbulence can seriously damage tower structures and turbine components. Turbulent installation sites must be avoided in the interest of preventing personal injuries and structural damage.

Mountains and ridges offer higher winds for reasons cited above other than channeling effects. Prominent peaks often pierce temperature inversions that can blanket valleys and low-lying plains. Temperature inversions cause stratification of the atmosphere near the surface and can affect wind environments. Normal air flow prevails above an inversion layer; below the layer the winds are stagnant—a condition detrimental to the efficient operation of a wind turbine. The air beneath an inversion layer may be completely cut off from air circulation within a weather system moving through the area. Temperature inversions are common in hilly and mountainous terrains such as Southern California and Western Pennsylvania. Both regions are notorious for their air pollution episodes. Inversions are most common during fall and winter in northern latitudes. Sometimes an inversion layer may accelerate the wind, thereby producing higher acceleration conditions which can create dynamic instability for a turbine tower structure.

2.5.3.2 Variations in Wind Speed and Direction

Wind speed and direction are subject to change over most measurement durations and under different weather seasons. The wind is an intermittent resource. The average wind speed can change as much as 25% from one year to the next. This can total more than 2 mph (1 m/sec) in a moderate wind regime where average wind speed of 10 mph (5 m/sec) is the norm. Generally the average wind speeds vary by season and by month. For example, early spring in March is windy, while the summer is not. For most interior regions in North America, winds are generally light during the summer and fall and increase during the winter, reaching the maximum in the spring. Accurate information on wind speed and direction gathered over 10 years or more should be reviewed by meteorologists before a site can be selected for a wind turbine installation.

It is interesting to note that increased wind speeds have been observed during late afternoons because of convective heating effects. Convective circulation leads to a dramatic difference between daylight and night wind speeds. Convective circulation is poor in winter. During winter and spring, winds are dominated by storm systems that can produce turbulent conditions. Meteorological data indicate that the average wind speed consistently exceeds 12 m/sec from April to August. This means one can expect more wind energy to be captured by a wind turbine during those months.

2.5.4 Fundamental Properties of Wind Energy

Performance capabilities absolutely depend on the properties of wind energy. Wind turbines operate on the theory of fluid mechanics [3] and therefore preliminary knowledge of the fundamental principles of fluid mechanics is necessary for thorough understanding of wind turbine operation. As noted in Chapter 1, the kinetic energy of the wind is used to drive turbines. As the blades of the wind turbine move through the air, the kinetic energy of the wind is converted into mechanical energy that drives the turbine coupled to an electric generator to produce electricity.

2.5.4.1 Available Wind Power

The kinetic energy of the wind (E_K) can be written as:

$$E_K = [(1/2)\ mV^2] \tag{2.4}$$

where m is the mass of the air and V is the wind speed (velocity). Although speed and velocity have distinct connotations and distinct meanings, they are often used synonymously. The expression for mass can be written as:

$$m = [\rho\ A\ V\ t] \tag{2.5}$$

where ρ is the air density (1.225 Kg/m³), A is the area through which the wind passes during a given period t, and V is the wind speed. Inserting Equation (2.5) into Equation (2.4), the expression for kinetic energy can be written as:

$$E_K = [(1/2)\ \rho\ A\ t\ V^3] \tag{2.6}$$

2.5.4.2 Power Generated from Wind Energy

A wind turbine designer is extremely interested in the amount of power he can generate from the wind energy available. Note power is defined as the rate of wind energy available per unit time. This can be also defined as the rate of wind energy passes through an area per unit time. Based on this definition, the power available from the wind is

$$P = [(1/2)\ \rho\ A\ V^3] \tag{2.7}$$

It is evident from Equation (2.7) that available wind turbine power is directly proportional to air density at the installation location and the area intercepted by the wind (swept area of blade). The available power is proportional to the cube of wind speed V. The swept area is be defined as:

$$A = [\pi\ R^2] \tag{2.8}$$

where A is the swept area of the turbine blade and R is the radius of the rotor, approximately equal to the length of the blade. The rotor blades spin about the horizontal axis and sweep a disc-shaped area equal to the area of a circle with a radius equal to R. Close examination of the parameters of Equations (2.7) and (2.8) reveals that significant wind power can be generated by deploying longer blades at installation sites where high winds are prevalent.

2.5.4.2.1 Impact of Air Density on Power from Wind

Meteorological data indicate that air density decreases with increasing temperature and increasing altitude. This means higher air density occurs at turbine sites located at lower altitudes with cooler temperatures. In other words, a place like the Gobi Desert, which is located at higher altitude but at lower altitude, is best suited for a wind turbine installation. That is why the Chinese installed several wind turbines capable of producing more than 10,000 MW annually in that region.

It is important to mention air density exerts negligible impacts on performance because most high-power wind turbines are installed at heights around 250 ft. Meteorological data reveal that air density changes little as installation height increases. The relative change in air density is about 99% at an installation height of 152 m, 97% at 305 m, and 94% at 610 m, assuming 100% air density at sea level. Note that air is less dense in summer than in winter. On a yearly average, seasonal changes in temperature produce negligible effects on the power available from wind. Furthermore, increasing altitude has more effect on air density than average temperature. For example, at the same wind speed, more wind power is available at a coastal site than a site in Denver, Colorado, located 5000 feet above sea level.

2.5.4.2.2 Impact of Swept Area on Power from Wind

Any variation in the swept area of a wind turbine will influence the power available from wind. Wind turbines with large rotors intercept more wind than those with smaller rotors and consequently capture more power from the wind. Doubling the area swept by a wind turbine rotor will double the power available from wind according to Equation (2.7). In other words, doubling the rotor radius will increase both the power from the wind and the swept area by a factor of four. This principle is fundamental to understanding the wind turbine design concept.

2.5.4.2.3 Impact of Wind Speed and Distribution on Power Available

According to Equation (2.7), power available from wind is proportional to the third power of the wind speed. Thus, if the wind speed is doubled, the power available will increase by a factor of eight ($2 \times 2 \times 2 = 8$). Even if the wind speed increases by 25%, a power increase of 95.3% is achieved. This indicates that even a slight change in the average annual wind speed can produce a significant boost in power from the wind. Clearly, wind turbine installation sites with higher average annual wind speeds must be given priority because of the availability of higher wind power levels throughout the year.

2.5.4.2.4 Rayleigh and Weibull Distributions for Estimating Wind Speeds

The distribution of winds at various speeds differs from one site to the next, but in general follows a bell-shaped curve for a family of Rayleigh distributions at sites with different wind speeds. Note the Rayleigh distribution is a hypothetical distribution of wind speeds based on a mathematical formula that tends to approximate the real world. It is interesting to mention that the power density (P/A) calculated from the Rayleigh distribution curve for a given average wind speed is almost twice that derived from the average wind speed alone. This relationship holds for many sites with moderate to strong average annual wind speeds. The Rayleigh distribution produces fairly reliable estimates when used to project annual power generation from small wind turbines. Some wind turbine project managers claim that their estimates using Rayleigh distribution were only about 5% less than the power actually produced. However, the monthly estimates were found less reliable. The Rayleigh distribution tends to overestimate power generation by 10 to 20%.

2.5.4.2.4.1 Rough Estimate of Power Density—Exact calculation or estimate of the power density from a wind turbine is very difficult because of variations in wind speed as a function of time. Under these conditions, an empirical formula defined by Equation (2.9) can provide a rough estimation of power density. Assuming a sea level air density of 1.225 Kg/m³ and a temperature of 60°F and inserting these values in Equations (2.7) and (2.8), the expression for the power density (P/A) in watts per square meter (W/m²) may be written as:

$$P/A = [(0.0547)\ V^3], \text{ where V is expressed in miles per hour (mph)} \quad (2.9)$$

$$P/A = [(0.6125)\ V^3], \text{ where V is expressed in meters per second (m/s)} \quad (2.10)$$

V is the average wind speed at sea level. Note these two equations are valid up to turbine heights close to 150 feet because the error in air density is less than 1%.

Wind speed distribution studies performed by the author indicate that speed distribution plays a very important role in determining the power from wind and it is always preferable to use an actual measured distribution at a specific site to achieve more accurate and reliable power estimates. Estimates prepared based on measured distribution and Rayleigh distribution curves can exhibit huge differences. For example, a measured distribution produces a power density of 320 W/m², twice the result from a Rayleigh distribution (160 W/m²) under the same average speed of 5.3 m/sec (12 mph).

Note the average annual wind speed defines the shape of the Rayleigh distribution curve. Close examination of this curve indicates that as the average annual wind speed increases, the curve shifts toward the higher wind speeds on the right side of the curve. The power density calculated from the Rayleigh distribution curve for a given average wind speed is almost twice that derived from the average wind speed alone. Rayleigh distribution remains a useful tool for most wind turbine sites, but meteorologists often prefer a more flexible mathematical formula called the Weibull distribution that more closely models wind turbine speeds at a wide range of installation sites. It is interesting that the Rayleigh distribution is a member of the Weibull family of speed distribution concepts.

2.5.4.2.5   Models for Calculating Power Density

Power density from a wind turbine can be calculated in two ways. One can sum a series of power density calculations for each wind speed and its frequency of occurrence (number of hours per year the wind blows at that speed) for the prospective site's distribution of wind speeds. One can also use the average wind speed and the cube factor for the Rayleigh distribution of 1.9. The relationship between the power density derived from the average speed alone and that from a speed distribution curve is commonly known as the cube factor or energy pattern factor. It is well established that the cube factor for the Rayleigh distribution is 1.9. Inserting the 1.9 cube factor into Equations (2.9) and (2.10), average annual power density (W/m²) can be estimated using the following modified equations:

$$\text{Average annual P/A} = [1.9(0.0547)] \ V3, \text{ where V is in mph} \qquad (2.11)$$

$$\text{Average annual P/A} = [1.9(0.6125)] \ V^3, \text{ where V is in m/s} \qquad (2.12)$$

Using Equations (2.11) and (2.12), one knows the average annual power density at a prospective site and can quickly estimate how much power a typical wind turbine can extract from the wind at that location. The impact of speed distribution on power density for various sites with the same average annual speed is evident from the data summarized in Table 2.4.

Note that site C is located in mountain passes subjected to funneling effects that cause more occurrences of winds at high speed than would be predicted by the Rayleigh distribution curve. As stated before, Rayleigh distribution at high-wind locations tends to overestimate power density by 10 to 20%. That is why the relevant cube factor is 2.4 instead of 1.9.

Table 2.4   Comparison of Impact of Speed Distribution on Power Density at Sites with Same Average Annual Wind Speeds

Site	Average Annual Wind Speed (m/s)	Power Density (W/m²)	Cube Factor
A	6.3	220	1.4
B	6.3	285	1.9
C	6.3	365	2.4

2.5.4.3 Impacts of Wind Speed and Installation Height on Performance

The combined impacts from wind speed and installation height are very complex. Brief studies of wind characteristics performed by the author indicate that wind speed and installation height above the ground affect the power generating capability of a wind turbine. In general, wind moving across the earth's surface encounters friction caused by turbulent flows over and around mountains, hills, trees, buildings and other obstructions in its path. These effects generally decrease as height above the ground increases until unhindered air flow is restored. In brief, wind speed increases as both the friction and turbulence decrease.

It is interesting to mention that frictional effects are strictly dependent on surface roughness. Friction effect is higher around tall trees and building structures than it is over the smooth surface of a lake or coastal region. In addition, the rate at which the wind speed increases with height varies with the degree of surface roughness. Wind speeds increase with height at a greater rate than wind speed rates over hilly and mountainous terrains. That is why taller towers offer higher wind speeds and permit the deployment of larger turbine blades, thereby yielding significant performance in terms of annual output power.

The studies further indicate that at low wind speeds, the change in wind speed as a function of height known as wind shear is less pronounced and more erratic, and thus detrimental to turbine performance. When temperature inversion occurs in calm wind environments, wind speeds may increase slightly between the ground and certain tower height and then begin to decrease. This means that the changes in wind speed with height are not constant. In certain mountain passes, wind speeds more than 200 ft (60 m) above the ground decreased with increasing height. However, when the average the wind shear is positive, wind speed increases with height. The following equation is useful for estimating the increase in wind speed as a function of tower height:

$$V/V_0 = [H_0/H]^\alpha \qquad (2.13)$$

Table 2.5   Changes in Wind Speed as Function of Height and Anticipated Value of Parameter α

Location	Speed at Height (m/s)		Increase in Speed	Parameter α
	30 ft	*150 ft*		
A	6.2	7.7	1.24	0.13
B	6.8	8.4	1.24	0.13
C	5.4	7.3	1.35	0.18
D	6.5	8.4	1.29	0.10
E	6.3	8.1	1.29	0.16

where V is the wind speed at the new height H, V_0 is the wind speed at the original height or the height of the anemometer attached to the turbine nacelle, and α is the frictional coefficient that varies with surface roughness and measures the friction encountered by the wind as it moves across the terrain. Changes in the wind speed as a function of height and frictional coefficient at various locations can be seen from the data summarized in Table 2.5.

Note that the increases in both the speed and the exponent of wind shear (α) occur at the installation location C, indicating that the ground obstructions have peculiar terrain footprints compared to other locations. Wind turbine installers claim that over smooth, level, grass-covered terrain, the parameter α or exponent for wind shear is very close to 0.14 or 1/7. When the exponent of 0.14 is applied to the ratio of two heights of 150 and 30 ft, the result is 1.253, indicating that the wind speed at the new height (H) of 150 ft will be 1.253 times that at the original height of 30 ft, or 25.3% greater.

A typical increase in wind speed as a function of height can be related to the anticipated increased surface roughness exponent or parameter α. The increase in wind speed as the installation height increases can be related to the surface roughness exponent as illustrated by Table 2.6.

2.5.4.3.1 Variations in Wind Shear Law

Although wind shear often follows the 1/7 (0.14) law, exceptions occur. Obstructions significantly reduce wind speeds near the ground. Furthermore, wind speed increases more dramatically with height than the 1/7 law predicts, particularly when the wind passes over corn crops, hedges, and scattered trees. In these situations, the wind shear could rise to 1/5 (0.20). When the surface is even rougher and the wind passes over more trees and few buildings, the wind shear exponent may increase to 1/4 (0.25).

Table 2.6   Increase in Wind Speed with Installation Height

Height		Surface Roughness Exponent			
(ft)	*(m)*				
20	6	0.10	0.14	0.20	0.25
30	9	1.00	1.00	1.00	1.00
60	18	1.07	1.10	1.15	1.19
80	24	1.10	1.1.5	1.22	1.28
100	30	1.13	1.19	1.27	1.35
120	37	1.15	1.22	1.32	1.41
150	46	1.17	1.26	1.38	1.50
160	49	1.18	1.27	1.40	1.52

2.5.4.3.2 Impact of Wind Shear Exponent on Output

It has been already demonstrated that the velocity at the new height will be equal to the wind shear exponent on the ratio of new height to initial height. The wind shear exponent used is equal to 0.14. Note that power is a cubic function of wind speed, as evident from Equation (2.7). If P_0 is the power at the original height of 30 ft and P is the power at the new turbine height of 150 ft, the ratio of these two powers based on Equation (2.13) can be written as:

$$P/P_0 = [H/H_0]^{3\alpha} \qquad (2.14)$$

where the wind shear exponent for this particular site α is equal to 0.14. This means that the power ratio for this site can be written as:

$$P/P_0 = [150/30]^{0.42} = [5]^{0.42} = [1.966]$$

Thus, the turbine output power at a height of 150 ft will be 1.966 times the power at 30 ft. Based on this sample calculation, one can expect greater power outputs at higher installation sites. That is why the wind turbines with outputs exceeding 2.5 MW require higher installation.

2.5.5 *Global Wind Power Capacity Using High-Power Turbines*

Based on a wind turbine manufacturers' survey, General Electric and Mitsubishi are deeply involved in the manufacture of high-capacity wind turbines. Present-day

Table 2.7 Global Wing Turbine Capacities (MW) at Higher Outputs

	Year of Operation	
Country	2005	2008
Germany	12,885	16,226
Spain	10,127	11,800
United States	9,125	11,975
India	4,435	6,125
Denmark	3,128	4,572
Italy	1,718	1,927
United Kingdom	1,353	1,512
China	1,260	1,689
Netherlands	1,219	1,482
Japan	1,040	1,331
Others	7,436	8,920
Total	59.088	72,533

high-capacity wind turbines operate in the range of 2.5 to 3.6 MW. Data on global wind turbines operating in the high- and medium-capacity ranges are summarized in Table 2.7.

The worldwide installed wind turbine capacity was 59,088 MW in 2005 [3]. The projected worldwide installed capacity increased to 72,533 MW during 2008. China, India, and Germany added more wind power capacity between 2005 and 2008 than other countries. The power capacity numbers for 2008 are based on realistic projections and good engineering judgment.

2.6 Annual Energy Acquisition from Specified Wind Turbine Site

The question of how much annual energy can be acquired from a specific wind turbine site must be answered before any decision to proceed is made. The answer to this question is a critical economic judgment. A realistic answer requires preliminary engineering judgment and an economic feasibility study for the specific site. A comprehensive study is required to estimate the value of the wind energy to be captured in a wind turbine lifetime relative to the initial investment, operating, and maintenance costs. The total wind energy available depends on the installation location and the annual variations in wind speeds and directions. Detailed data on the distribution of wind speeds and directions over a period of several years at the specific site must be examined to allow accurate long-term predictions of wind energy capture. If very large and expensive wind turbines are involved, the design, development, installation, and operating costs warrant detailed long-term wind surveys. In brief, cost-effective operation and high wind conditions are critical considerations for choosing an installation site.

2.6.1 Requirements for Long-Term Capture of Wind Energy

Appreciable annual wind energy acquisition requires high wind conditions 24 hours a day every day. This is the most fundamental requirement for selecting an installation location. Turbine designers have estimated wind turbine power output levels for a 125-ft diameter turbine rotor as a function of annual wind speed. The estimated power output levels are depicted in Table 2.8 and demonstrate a trend in turbine output power with increases in wind speed for that specific site and rotor size.

2.6.2 Impact of Wind Speed on Wind Energy Density

If a wind energy density in kilowatt hours per square foot per year (kWh/ft²/year) is plotted as a function of wind speed (mph), one will find a bell-shaped curve showing the maximum energy content wind region in the middle of the curve and the low energy content regions at the ends of the curve. This curve was plotted for wind speeds from 0 to 70 mph. The maximum energy-content region for this particular wind turbine with 100-kW capacity extends from 18 to 25 mph, while the minimum energy content regions extend from 0 to 5 mph and from 45 to 70 mph. However, intermediate energy content regions exist between the maximum

Table 2.8   Estimated Power Output from Turbine with 125-Foot Diameter Rotor

Wind Speed (mph)	Wind Turbine Power Output (kW)
10	22
15	65
20	185
25	375
30	596

and minimum regions. Note these parameters apply only to a 100-kW system with a rotor diameter of 125 ft. Energy content regions for wind turbines will exhibit different wind speed ranges compatible with characteristics of the bell-shaped curve of wind energy density versus annual wind speed for high-power turbines and other installation locations.

2.6.3 Annual and Hourly Extraction of Wind Energy by Wind Turbine

It is extremely difficult to estimate the wind energy that can be captured per hour or per year when a wide range of wind conditions is encountered. Furthermore, the unsteady nature of the wind involves the transient responses of the wind turbine system that cannot be determined from the time integrals of wind speeds and turbine power outputs. The total wind energy available per hour or per year depends on the installation site, fluctuations in wind speeds, unpredicted variations in wind directions, and structures, trees and surface conditions in the vicinity of the turbine site.

Wind turbine designers claim that the highest wind energy capture per hour or per year can be achieved when the speed at which power limiting is initiated is about 50% of the average wind speed. However, this criterion is not cost-effective because a turbine designed for such low speed would have to be very large and expensive for its nominal power output. Note that calculations of the energy integrals as a function of wind speed ratio provide useful information on wind energy captured per hour. The energy integral is defined as the ratio of wind energy captured at final wind speed (E) to the wind energy captured at initial wind speed (E_0) and it varies from rotor to rotor [1].

2.6.4 Energy Integrals for Savonius Rotor in VAWT

Calculations of energy integrals for Savonius rotors widely used for VAWTs can be performed using the following equation:

$$F = [4 \, Y \, z \, (1 - Yz)] \tag{2.15}$$

where z is equal to $1/\lambda$, Y is a constant (its value varies from 0.6 to 0.7 for a Savonius rotor), and λ is the wind speed ratio varying from 1 to 3 for this particular rotor. It is extremely important to mention that the constant Y will have a very limited range of values because the upper and lower portions of the curve represent extreme wind conditions that cannot be tolerated by this particular rotor.

The energy ratio (E/E_0) curve as a function of wind speed ratio (λ) can be determined for a given rotor configuration and is used to indicate whether an energy ratio near to or greater than unity can be achieved for Savonius turbine rotors. At an average wind speed of 9.5 mph or 4.25 m/sec in the U.S., the total annual

Table 2.9 Energy Integral as Function of Wind Speed Ratio for Savonius Rotor

Wind Speed Ratio (λ)	Values of Constants		
	0.6	0.7	0.8
1.0	0.168	0.122	0.102
1.5	0.732	0.692	0.654
2.0	1.242	1.214	1.205
2.5	1.445	1.427	1.393
3.0	1.474	1.425	1.4

energy flow is around 38.6 kWh/ft^2 of swept area. Assuming a maximum power coefficient of 0.3 (typical for a Savonius rotor), the total annual wind energy capture by a wind turbine would be about 11.6 kWh/ft^2 of swept area. The energy integral (E/E_0) values for a Savonius rotor as a function of wind speed ratio (λ) are shown in Table 2.9.

Similarly values of the energy integrals can be obtained for other types such as upwind or downwind rotors deployed by HAWTs. Since the maximum power coefficient for a HAWT is close to 0.59 as illustrated in Figure 2.7, the total annual wind energy captured by a HAWT would be 1.967 times more than that achieved by a Savonius rotor with the same average wind speed of 9.5 mph. In brief, the HAWT would capture total annual wind energy close to 22.77 kWh/ft^2 of swept area.

2.6.5 Use of Vortices for Creating Regions of High Wind Velocity

As noted, higher wind speed of velocity will cause a wind turbine to generate more power. Higher wind speeds can be achieved by increasing tower heights or using augmentation techniques incorporating aerodynamic surfaces or vortices generated by buildings, wind tips, or delta wings. A more promising approach seems to be the use of tip vanes—aerodynamic surfaces placed on the tips of the blades at such an angle that their lift tends to turn the wind into the rotor, thus increasing wind speed and enhancing the power output capability of the turbine. In brief, this is considered as an approach to augmenting the power output of a rotor. Augmentation concepts for rotors will be discussed in detail in Chapter 3.

The approach of using the ejectors or ejector–diffusers incorporating aerodynamic surfaces has been most successful for augmenting the power output of a turbine by increasing the pressure drop across the rotor as shown in Figure 2.9. The potential effectiveness of such augmentation elements can be easily estimated when the Bernoulli equation is written for the flow ahead of the rotor and then for the downstream flow in the same form. The Bernoulli equations for both flow conditions can be written as:

$$[P_0 + PV_0^2/2] = [P' + \rho V'^2/2] \tag{2.16a}$$

$$[P_1 + \rho (V_0 - q)^2/2] = [P'' + \rho V'^2/2] \tag{2.16b}$$

where P_0, V_0 and P_1, V' are the pressure and wind velocities very far ahead of and very far behind the rotor, respectively. The velocity deficit at the location of P_1 is defined by the parameter q. The velocity of the rotor is denoted by V'. The parameters P' and P'' are the pressures immediately ahead of and immediately behind the rotor, respectively. When these two Bernoulli equations are solved for the pressure drop without the assumption of P_0 equal to P_1, The result is:

$$[P' - P''] = (\rho V_0^2/2) [(2 q/V_0) (1 - q/2 V_0) - c_{pr}] \tag{2.17}$$

where c_{pr} is the exit pressure coefficient across the rotor. Its expression can be written as:

$$C_{pr} = [(P_1 - P_0)]/[\rho V_0^2/2] \tag{2.18}$$

2.6.6 Maximum Power Coefficient as Function of Exit Pressure Coefficient and Interference Factor (a)

Power is defined as the rate of work done by the fluid moving through the rotor. Computations of power coefficient require appropriate values of interference factor (a). However, the maximum power coefficient occurs at a value of interference factor a whose empirical expression can be written as:

$$a = [2/3 - 1/3 (1 + c_{pr})^2] = [0.666 - 0.333 (1 + c_{pr})^2] \tag{2.19}$$

where c_{pr} is the exit pressure coefficient. Computed values of the interference factor a as a function of exit pressure coefficient are depicted in Table 2.10.

2.6.7 Computations of Power Coefficients

Power coefficients are functions of exit pressure coefficients and interference factors. The expression for calculating power coefficient can be written as:

$$C_p = [4a (1 - a)^2 - (1 - a) c_{pr}] \tag{2.20}$$

Table 2.10 Computed Values of Interference Factor as Function of Exit Pressure Coefficient

Exit Pressure Coefficient (c_{pr})	Interference Factor (a)
00	0.333
0.1	0.266
0.2	0.186
0.3	0.103
0.4	0.014

By inserting various values of interference factor a, power coefficient can be computed as a function of exit pressure coefficient. Computed values of power coefficients are summarized in Table 2.11.

2.6.7.1 Techniques for Enhancing Power Output and Power Coefficient

Certain potential augmentation techniques using aerodynamic surfaces may be deployed to enhance the power output levels and the power coefficients of wind turbines. Techniques involving ejectors for the wind turbines must be first analyzed in terms of energy and momentum balances. High potential gains from the use of diffuser–ejector schemes may be illusory if a complete wind turbine system study is not performed. A comprehensive study can be performed within a shroud that acts like an ejector with a constant area mixing duct followed by a diffuser. Studies performed by various wind turbine scientists [4] indicate that the power outputs of wind turbines could be augmented by a factor of 10 to 20, but the power coefficient referenced to the enormous area was inferior to that of a free turbine system. The enormous size and additional weight of an augmentation device could be justified.

Table 2.11 Wind Turbine Power Coefficients

Exit Pressure Coefficient (c_{pr})	Interference Factor (a)			
	0.4	0.3	0.2	0.1
00	0.576	0.588	0.512	0.324
0.1	0.516	0.518	0.432	0.234
0.2	0.456	0.448	0.332	0.144
0.3	0.396	0.378	0.272	0.054
0.4	0.336	0.308	0.215	0.008

However, this particular augmentation technique confines the turbine installation to ground level, thus reducing the kinetic energy available from the wind due to lower wind speeds.

2.7 Estimating Annual Hours of Capturing Wind Energy

Preliminary estimation of annual hours that could be expected to capture the wind energy at a potential installation site is critical for justifying the capital investment and ensuring cost-effective operation. Annual hours to capture the wind energy from a particular site are based on wind speed ratio and other variables.

2.7.1 Annual Hours Estimation Using Empirical Method

The annual hours to capture wind energy can be calculated from the following empirical expression:

$$H(\lambda) = [a\ b^{bz(\lambda)}] \tag{2.21}$$

where λ is the wind speed ratio that typically varies from 1 to 3 for maximum hours, a represents the annual hours (8760 = 365 days × 24 hours), z (λ) is a function of wind speed and is equal to λ^c, b has a typical value of 0.7738 for maximum hours, and c values range from 1 to 5. Inserting a fixed value of 8760 for a, 0.7738 for b, and various values for c in Equation (2.21) allows annual hours to be calculated. Table 2.12 shows calculated annual hours for capturing wind energy.

Wind energy captured hours have been plotted as a function of wind speed and various values of parameter c. Figure 2.10(a) shows annual hours when parameter c equals 1; Figure 2.10(b) shows annual hours when c equals 2; Figure 2.10(c) shows annual hours when c equals 2.27; Figure 2.10(d) shows annual hours when c equals

Table 2.12 Annual Hours for Capturing Wind Energy

Wind Speed Ratio (λ)	Values of Parameter c				
	1	2	3	4	5
0	8760	8760	8760	8760	8760
1	7188	7188	7188	7188	7188
2	5894	3964	1792	366	15
3	4834	1470	41	00	00

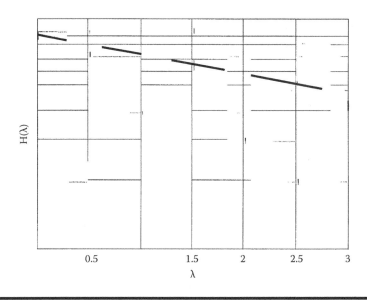

Figure 2.10(a)   Plot of annual hours as function of wind speed ratio when factor c is: (a) equal to 1.

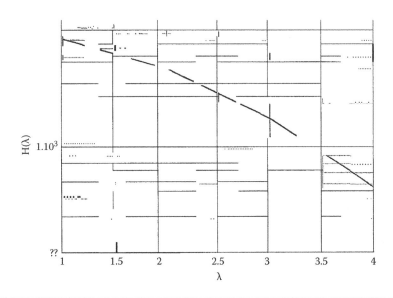

Figure 2.10(b) (continued)   Plot of annual hours as function of wind speed ratio when factor c is: (b) equal to 2.

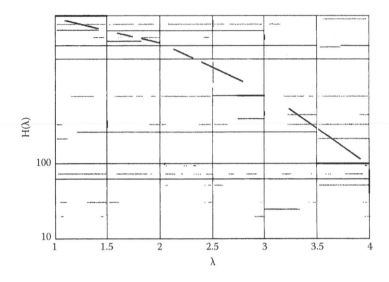

Figure 2.10(c) (continued)   Plot of annual hours as function of wind speed ratio when factor c is: (c) equal to 2.27.

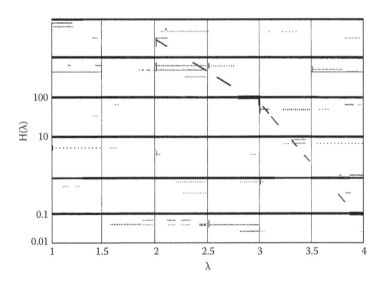

Figure 2.10(d) (continued)   Plot of annual hours as function of wind speed ratio when factor c is: (d) equal to 3.

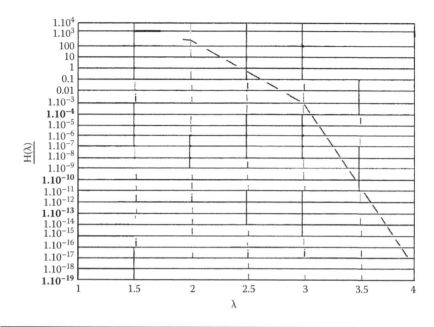

Figure 2.10(e) (continued) **Plot of annual hours as function of wind speed ratio when factor c is: (e) equal to 4.**

3; Figure 2.10(e) shows annual hours when c equals 4; and Figure 2.10(e) shows annual hours when c equals 5. We can conclude from these plots that maximum hours to capture wind energy are possible when parameter c varies from 1 to 2.27 at wind speed ratios between 1 and 2. In summary, the principal objective of a wind turbine installer is to select an installation site that offers maximum annual hours for capturing wind energy over a wind speed ratio between 1 and 2.

The plots indicate that the maximum duration of annual hours as a function of wind speed ratio can be achieved when parameter b is unity instead of 0.7738 and parameter c equals 2 as in Figure 2.10(g). Note these plots merely provide preliminary estimates for the annual hours over which wind energy can be captured.

2.7.2 Annual Energy Production Estimate Using Blade Element Momentum Method

It is possible to generate a power curve using the blade element momentum (BEM) method to estimate shaft power as a function of wind speed V_0. To compute annual wind energy production, it is necessary to combine the production curve with a probability density function of the wind. From this function the probability $[f (V_i < V_0 < V_{i+1})]$ that the wind speed lies between V_i and V_{i+1} can be computed.

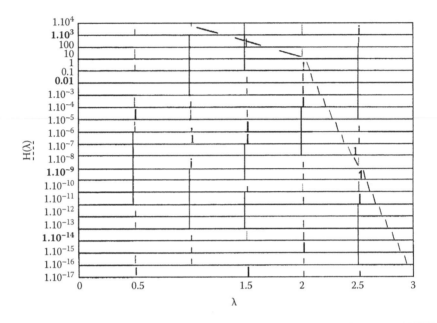

Figure 2.10(f) (continued) Plot of annual hours as function of wind speed ratio when factor c is: (f) equal to 5.

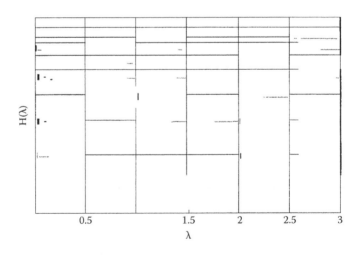

Figure 2.10(g) (continued) Plot of annual hours as function of wind speed ratio when factor c is: (g) equal to 2.

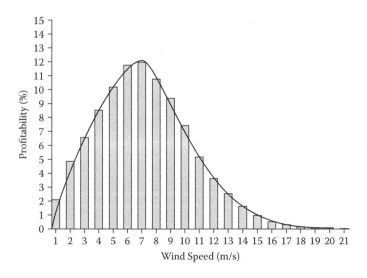

Figure 2.11 Wind speed defined by widths of columns. Histogram and Weibull distribution function.

Multiplying this with the total hours per year (8760) yields the number of hours per year that the wind speed lies in the interval $V_i < V_0 < V_{i+1}$. Multiplying the probability function by the power (in kilowatt hours) produced by the wind speeds ranging from V_i to V_{i+1}, one can determine the contribution of total production (in kilowatt hours) for this speed interval. The wind speed can be discretized into N discrete values $(V_i, I = 1, N)$, with a typical speed interval of 1 m/sec. In brief, a probability density function that the wind speed lies between V_i and V_{i+1} and a shaft power curve are needed to compute annual energy production for a specific wind turbine at a particular site.

Note that wind energy production must be corrected for the losses at the generator and gear box. The combined efficiency of these two components is approximately 90%. The probability density function of the wind is determined by a Rayleigh or Weibull distribution. It is important to mention that the Weibull offers a more accurate probability density function because some corrections for landscape, vegetation, and structures can be modeled through a scaling factor A and form factor k. From the Weibull distribution, the probability density function for the wind can be written as:

$$f(V_i < V_0 < V_{i+1}) = [\exp(-(V_i/A)^k)] - [\exp(-(V_{i+1}/A)^k)] \qquad (2.22)$$

The total annual energy production (AEP) can be calculated as:

$$\text{AEP} = [\sum_{I=1}^{N-1}(1/2) \, P \, (V_{i+1}) + P \, (V_i) \times f \, (V_i < V_0 < V_{i+1}) \times 8760] \qquad (2.23)$$

Mechanical shaft power is a function of wind speed, air density, and relevant parameters of the airfoil. The data available in the literature specify an airfoil not thicker than 20% of the chord and moderate angles of attack. To maintain structural integrity of the rotor blades, it is desirable to use a very thick airfoil of approximately 40% of the chord at the roots of each blade to absorb the high bending moments. Furthermore, the secondary layers on the rotating blades are affected by the centrifugal and Coriolis forces that alter the post-stall lift and drag coefficients from those measured in wind tunnel tests.

It is interesting to mention that the Coriolis acceleration due to the rotation of the earth causes the flow from the equator to the poles that affects wind parameters and leads to the formation of trade winds. Clearly, significant engineering skill and design and development experience are necessary to calculate reliable data for thick airfoils with high angles of attack including three-dimensional factors such as the rotational effect. When the power curve showing the actual wind turbine shaft power (kilowatts) as a function of wind speed V_0 (meters per second) is obtained from the measurements, it is then desirable to calibrate the airfoil data to achieve better agreement between the measured and computational data. If a new blade very similar to the blades for which the airfoil data were calibrated is fabricated, it is possible to predict the power curve with high accuracy. However, if a new blade is designed with a completely new airfoil, the computed results should be used carefully. In brief, the actual geometry and airfoil data for the blade are required to extrapolate the data for a high angle of attack. Typical turbine mechanical shaft power (kilowatt) levels as a function of wind speed (meters per second) are summarized in Table 2.13.

Based on these tabulated shaft power data, except at very low wind speeds, the BEM method provides turbine shaft power levels with minimum differences between measured and calculated data collected on a wind turbine having the following design parameters:

- Rotor radius = 20.5 m
- Rotational speed = 27.1 rpm

Table 2.13   Mechanical Shaft Power as Function of Wind Speed

Wind Speed, V_0 (m/sec)	Turbine Shaft Power (kW)
5	42
10	318
15	516
20	562
25	570

- Number of blades = 3
- Hub height = 35 m
- Air density = 1.225 Kg/m^3

2.7.3 Parameters Affecting Performance

Wind turbine blades designed for optimum performance require variations of the airfoil sections along the radius of the blade to ensure that each blade element operates at its minimum drag-to-lift ratio (ε) that represents the ratio of drag coefficient to lift coefficient. The tangential speed ratio (X) also plays a key role in the improvement of the power coefficient and is equal to $R\omega/V_0$, where V_0 is the wind speed, R is the blade radius, and ω is the angular velocity of the rotor blade. Variations of the power coefficient and axial interference factor (a) as a function of tangential speed ratio are evident from the data summarized in Table 2.14. Variations in power coefficient as functions of various parameters will be discussed in Chapter 4.

2.7.3.1 Impact of Wind Characteristics on Power Coefficient

The amount of wind energy that can be captured depends on wind characteristics. As noted earlier, the power of the wind is proportional to the cube of the wind speed or velocity. It is therefore essential to have detailed knowledge of the wind characteristics if the performance of a wind turbine is to be estimated accurately. According to meteorological statistics, the highest wind speeds are generally found at hill tops, coastal regions, and out at sea. Various parameters such as mean wind speed, directional data, short-term variations (gusts), daily, seasonal, and annual variations, and wind parameter changes as functions of height must be known for a

Table 2.14　Impacts of Drag-to-Lift Coefficient and Tangential Speed Ratio on Power Coefficient (C$_p$)

Drag-to-Lift Ratio (ε)	Tangential Speed Ratio (X)							
	1	2	3	4	5	6	7	8
0.00	0.52	0.56	0.57	0.58	0.585	0.590	0.592	0.594
0.01	0.50	0.55	0.56	0.55	0.542	0.535	0.527	0.52
0.03	0.46	0.47	0.45	0.42	0.371	0.4.32	0.406	0.377
0.05	0.45	0.46	0.45	0.41	0.372	0.341	0.295	0.264
0.07	0.43	0.44	0.39	0.34	0.314	0.224	0.174	0.142
0.10	0.41	0.38	0.32	0.25	0.185	0.128	0.068	0.012

specific installation site. These wind parameters must be determined with sufficient accuracy through measurements over a sufficiently long period to accumulate reliable wind characteristic data that are essential to determine the performance and economics of a wind turbine plant.

2.7.3.1.1 Effects of Atmospheric Boundary Layer on Performance

Microflow is influenced by local features such as trees, structures, and terrain features in the vicinity of an installation site. The water surface of a lake or river or sea exerts important effects on wind characteristics. The resulting changes in friction at the surface produce variations in wind velocity. Also, wind velocity varies over time and space. An anemometer located at the nacelle of a turbine is capable of recording wind velocity data with great accuracy. Some variation in wind velocity arises from turbulence created by the wind turbine rotor and nacelle.

The expression for the instantaneous wind speed can be written as:

$$V = [V_m + dv] \qquad (2.24)$$

where V_m is the mean wind speed typically determined over a 10-min interval over a long period and dv is the fluctuating wind speed component. The fluctuation of the flow can be expressed in terms of the root mean square (RMS) value of the mean fluctuating wind velocity component (dv^2). This fluctuation velocity component is defined as turbulence intensity and can be written as:

$$T_{int} = [\text{RMS value of } dv^2/V_m] \qquad (2.25)$$

According To wind turbine installers, the turbulence intensity for very rough terrain consisting of trees and buildings is about 0.15 to 0.20.

2.7.3.1.2 Impact of Vertical Wind Speed Gradient on Power Coefficient

The vertical wind speed gradient can affect the power coefficient of a wind turbine. Note the wind speed at the surface is zero due to the friction between the air and ground surface. The wind speeds increases with height most rapidly near the ground and less rapidly at greater heights. At about 2 km above the ground, the change in wind speed becomes zero. The vertical variations in wind speed and wind speed profile can be expressed by appropriate functions. The change in mean wind speed with height in the form of a power exponent function can be written as:

$$V(z) = V_r [(z/z_r)^\alpha] \qquad (2.26)$$

where z is the height of the turbine above the ground, V_r is wind speed at the reference height at turbine height z_r above the ground, and α is an exponent that depends on the terrain roughness whose typical value varies from 0.10 to 0.15.

Sample Calculation:

If wind speed at 10 m height = 20 m/s, what will be the wind speed at 40 m height?

Wind speed at 40 m height $[20\ (40/10)^{0.1}] = [20 \times 1.1487] = 22.974$ or ~23 m/sec

2.7.3.1.3 Value of Exponent Parameter as Function of Roughness Class and Length

The values of the exponent parameters are strictly dependent on surface conditions of the terrain in the vicinity of the wind turbine installation sites. Note exact values are difficult to predict because of significant variations in surface roughness from calms and turbulent winds. Computed values of this exponent with an error of ±5% are summarized in Table 2.15.

2.7.3.1.4 Impact of Temperature Gradient on Wind Speed

The major factors affecting wind speed have been considered. Now the strong influence of the vertical temperature gradient on the speed and the speed difference will be discussed. It is customary in meteorology to correlate wind speed measurements using the following equation:

$$V_h = [V_1\ (h/h_1)^m] \qquad (2.27)$$

Table 2.15 Values of Exponent Parameter as Function of Surface Roughness

Terrain Type	Roughness Class	Roughness Length (m)	Roughness Exponent (α)
Water surface or area	0	0.001	0.01
Open countryside	1	0.12	0.12
Farm land with buildings	2	0.05	0.16
Village	3	0.29	0.28

Table 2.16   Reduction of Average Power Output by Wind Speed Gradient

Exponent (m)	Parameter (r/h₀)				
	0.00	*0.25*	*0.50*	*0.75*	*1.00*
1/7	1.000	0.990	0.983	0.958	0.897
1/5	1.000	0.996	0.984	0.960	0.910

where V_h is the horizontal component of the wind velocity at height h, V_1 is the horizontal component at height h_1, and m is the exponent that varies approximately between 1/5 and 1/7. This exponent is depends on local atmospheric conditions. Meteorology scientists have noted the strong influence of the vertical temperature gradient and also indicate that the vertical temperature gradient affects the wind speed gradient, thus reducing turbine average output power by a reduction factor R as illustrated in Table 2.16. R is defined as the ratio of average power output of a turbine to nominal power due to wind speed gradient. This ratio is calculated using the following equation:

$$R = (1/2\pi) \int [(1 + (r/h_0) (\sin \theta)]^{3m} d\theta \text{ with integration limits between } 2\pi \text{ and } 0. \quad (2.28)$$

where r is the blade element at radius R on the arm whose axis is located at height h_0 and the arm makes angle θ with the horizontal axis. It is important to mention that the tips of the rotor blades of a propeller turbine would not be permitted to operate very far below the r/h_0 ratio equal to 0.5. Under these conditions, the influence of the wind speed gradient observed is insignificant.

2.7.3.1.5 Wind Statistics

Histograms of wind velocity covering several years are available and these data allow the probability density function to be determined. This function can demonstrate where the wind speed falls within the interval defined by the widths of the columns as illustrated in Figure 2.11. The figure shows the histogram and Weibull distribution function of the probability for a given speed. Speed data are measured in meters per second "bins" such as 3.5 to 4.5, 4.5 to 5.5, 5.5 to 6.5, etc. The sum of the height of the columns is generally assumed as unity or 100%.

When the width of a column becomes very small, the histogram becomes a continuous function known as the probability density function that indicates the probability of a wind speed at a 1 m/sec interval centered on the value of the wind speed V. It is evident from the curve in Figure 2.11 that the probability of a wind speed between 4.5 and 5.5 m/sec is approximately 10.4% or 0.104. In other words, the probability of the wind offers annual energy (0.104 × 8760) of 910 hours. Note the histogram accounts for seasonal and annual variations for the years covered by the statistics.

The probability density function can be fitted to a Weibull distribution function written as:

$$P\,(V) = [(k/V)\,(V/C)^{k-1}]\,[\exp\{-\,(V/C)^k\}] \qquad (2.29)$$

P (V) is the frequency of occurrence function of the wind speed V, C is the scale parameter of the wind speed, and k is the shape parameter. A typical value for k is close to 2 and when k equals 2, the result is called a cumulative Rayleigh distribution. If k has a value other than 2, the result is known as a cumulative Weibull distribution. It is important to note that shape parameter k could be equal to 10 m/s or 5 m/s, regardless of Rayleigh or Weibull distribution. However, when shape parameter k is equal to 5 m/s parameter, percentage probability is much accurate.

2.7.4 Combined Impact of Rotor Tip Speed Ratio and Drag-to-Lift Ratio on Power Coefficient

The impact of tip speed ratio and drag-to-lift coefficient ratio on the turbine power coefficient as a function of angle of attack is a serious issue. At higher angles of attack, the power coefficient deceases rapidly, regardless of values of the other two parameters. The author's studies of angle of attack reveal that low angles are necessary for higher turbine outputs and power coefficients. The design and development of rotor blades with low angles of attack is a complex endeavor. Trade-off studies of rotor blades performed by the author indicate that an angle of attack equal to 0.1 radian or 5.73 degrees must be selected for blade design if maximum power coefficients are desired over a wide range of tip speed ratios. Estimated values of power coefficients for a turbine rotor blade with an angle of attack equal to 0.1 radian [1] for various tip speed ratios and drag-to-lift coefficient ratios are shown in Table 2.17.

It should be noted that peak values of the power coefficient tend to occur at lower values of tangential speed ratio as the drag-to-lift coefficient ratio (ε) increases. Rotor blade designers claim that very efficient blade sections with drag-to-lift coefficient ratios equal to 0.01 demonstrate a maximum when X is approximately equal to 3. Similar trends have been observed for variations of the axial interference factor. These variations will be discussed in Chapter 4.

These facts can be verified from the calculated data summarized in the previous tables. It is interesting to mention that the drag-to-lift coefficient ratio effects both turbine power coefficients and variations in rotor torque grading coefficients and blade loading coefficients as a function of tangential speed ratio. The impacts

Table 2.17 Power Coefficient (C_p) for Rotor as Function of Tip Speed Ratio and Drag-to-Lift Coefficient Ratio at Optimum Angle of Attack (0.1 Radian)

Tip Speed Ratio (λ)	Drag-to-Lift Coefficient Ratio (ε)		
	0.01	0.03	0.05
1			
2	0.38	0.46	0.44
3	0.53	0.51	0.47
4	0.55	0.50	0.43
5	0.56	0.45	0.27
6	0.56	0.34	000
7	0.53	000	000
8	0.50	000	000
9	0.44	000	000
10	0.37	000	000

of these variations on various rotor and blade performance parameters will be discussed in detail in Chapters 3 and 4.

Calculated values of power coefficients never exceed 0.59 because this value is limited by the mathematical derivation of the Betz limit. Thus the maximum resulting value of the power coefficient or coefficient of turbine performance theoretically is 0.59. This is the main reason that none of the power coefficient values summarized in the above table approached or exceeded 0.59. The Betz limit was formulated in 1919 and it applies to all types of wind turbines. It is critical to mention that if the air is brought completely to rest, all its energy will dissipate. Furthermore, a rotating wind turbine will not completely prevent the flow of air, so it can only extract a portion of the kinetic energy from wind. The wind speed onto the rotor at which the energy extraction is a maximum and the wind speed will lie somewhere between the windstream speed and zero wind speed.

In practice, most modern wind turbines deployed for electricity generation operate at power coefficient (C_p) levels close to 0.4. The reduction in the power coefficient limited by the Betz is due to the losses from the viscous drag on the rotor blades, the swirl imparted to the airflow by the rotor structure, losses at the gear box, and the electrical power losses in the transmission line, electrical generator, and step-up transformer at the tower base.

2.8 Summary

This chapter describes the various types of wind turbines with emphasis on their fundamental operating principles, critical design aspects, and performance capabilities and/or limitations. Performance capabilities of modern HAWTs and VAWTs are summarized. Advantages and disadvantages of both types of wind turbines are briefly mentioned. Potential rotor blade articulation mechanisms to regulate the angular rate of a rotor are identified with emphasis on reliability and dynamic stability of the turbine. Estimated values of power coefficients for HAWTs and VAWTs are provided.

Impacts of drag-to-lift coefficient ratios on maximum power coefficients as functions of wind tip ratios were discussed. Potential augmentation techniques involving aerodynamic lift surfaces, ejector ducts, and diffuser ducts to enhance performance were summarized, along with information about penalties due to additional weight, size and cost. Practical techniques to capture large amounts of wind energy are outlined with emphasis on reliability and turbine dynamic stability. The impacts of blade dimensions, pitch angles, and angles of attack on performance along with details about cost and design complexity are covered.

Wind characteristic requirements to enhance turbine performance are identified in relation to installation site environments. Computed values of wind turbine power output, power density, and energy available as a function of rotor diameter, installation site height, and swept area are provided. Sample calculations of wind turbine power output levels in relation to installation height, wind speed, and wind distribution are summarized. Important applications of Rayleigh and Weibull distribution functions to estimate the long-term average wind speed for a specific installation site are identified; annual wind energy capture is discussed. Wind energy models are briefly discussed in terms of wind speed and installation terrain. The effects of wind speed and installation height on performance are identified with emphasis on turbine output, dynamic stability, and design complexities. The impact of wind shear on turbine performance as a function of wind characteristics and surface roughness factors is briefly discussed.

Worldwide data concerning installed wind power capacity from 2005 through 2008 are summarized. The accuracy of these data is within ±5%. Energy integrals capable of providing power coefficients for wind turbines as a function of wind speed ratios are identified and the impacts of exit pressure and interference factor on maximum power coefficient are explained. Estimates of annual hours to capture wind energy from a specific installation site and as a function of wind speed ratio are provided. Annual wind energy production estimates using the blade element momentum (BEM) method are summarized.

Computed values of turbine mechanical shaft power as a function of wind speed are provided. Performance capabilities of propeller-based wind turbines as a function of drag-to-lift coefficient ratios and tangential speed ratios are briefly mentioned. The effect of vertical wind speed gradient on the wind turbine power coefficient is highlighted in relation to turbine height and terrain roughness. Reduction of average power output as a function of wind speed and wind speed gradients is discussed. Histograms of wind speed data collected over a long period play a critical role in providing reliable information on wind speed and direction for a specific installation site. Performance parameters based on tip speed ratios and drag-to-lift coefficient ratios are summarized with emphasis on reliability and maintenance-free operation.

References

[1] Europe replaces old wind farms, *IEEE Spectrum*, Jan. 2009, p. 13.
[2] A.J. Wortman, *Introduction to Wind Turbine Engineering*, Butterworth, Boston, 1983, p. 37.
[3] M.O.L. Hansen, *Aerodynamics of Wind Turbines*, 2nd ed., Jane Science, 1938, London, p. 18.
[4] T. Wizelius, *Developing Wind Turbine Projects*, Earth Scan Publishing, Sterling, VA, 2007, p. 107.

Chapter 3

Design Aspects and Performance Capabilities of Wind Turbine Rotors

3.1 Introduction

This chapter is dedicated to wind turbine rotors and associated components, with emphasis on rotor classification, design aspects, performance capabilities, and critical design requirements. Various types of rotors widely used for modern wind turbines will be described, particularly with regard to design simplicity and reliability. One-dimensional (1-D), two-dimensional (2-D), and three-dimensional (3-D) aerodynamic analyses will be undertaken on wind turbine rotors with a focus on power and thrust coefficients [1].

Initially, a simple 1-D model for an ideal rotor design configuration will be evaluated to identify the critical roles of axial velocity, pressure drops across the rotor, and the axial component of the pressure forces acting on the control volume. Critical circular control flow parameters will be identified along with their effects on axial velocity. Reasons for velocity jumps in the wake will be cited. Conditions and parameters required for stationary, incompressible, and frictionless flow will be identified. Use of the Bernoulli equation will be justified for the far upstream area to just in front of the rotor and from just behind the rotor. The axial momentum equation for an ideal rotor will be derived using certain assumptions. The expression for the rotational speed of a rotor will be developed as a function of axial induction and axial wind velocity.

3.1.1 Rotor Types and Their Performance Capabilities

Wind turbine rotors and their performance capabilities and critical performance parameters will be briefly described. Upwind and downwind rotors are widely used by HAWTs while the Savonius and Darrieus rotors are best suited for VAWTs. The efficiency of the Savonius rotor is generally poor over tip speeds ranging from 1 to 2.5, but it improves when the tip speed ratio exceeds 3. The rotor efficiency is about 53 and 57 at tip speed ratios of 3 and 4, respectively. The Savonius rotor has demonstrated a constant efficiency better than 58% over tip speed ratios between 5 and 10. Tip speed ratio is defined as the ratio of blade tip speed to undisturbed wind speed.

The efficiency of a Darrieus rotor varies roughly from 28 to 32% over tip speed ratios from 5 to 7, with a maximum efficiency of 33% at a tip speed ratio of 6. Because of its relatively poor efficiency, the deployment of the Darrieus rotor is restricted to low-capacity wind turbines.

Upwind and downwind rotors have demonstrated higher efficiencies and are generally deployed by HAWTs and propeller type windmills. The ideal efficiency for rotors on windmills is high: 41.5, 51.2, 54.8, 57.2, 57.5, and 58% at tip speed ratios of 1, 2, 3, 4, 5, 6 and 7, respectively. Both upwind and downwind rotors are best suited for high-capacity turbines operating at high tip speed ratios.

3.1.1.1 Rotor Blades

The propeller blade is the most critical element of a wind turbine rotor. Considerable importance is attached to performance, safety, and mechanical integrity of a rotor during its design phase. The rotor blades must follow all the principles of aerodynamic and fluid mechanics to achieve optimum performance regardless of wind environment. Design configurations and performance requirements for the rotor blades are different for HAWTs and VAWTs. Preliminary studies performed by the author indicate that the maximum efficiency or maximum power coefficient is possible at increasing values of tip speed ratios as the drag-to-lift coefficient ratio diminishes. The studies further indicate that all the blade sections must be pitched to ensure a minimum drag-to-lift ratio. Design configurations and performance requirements for rotor blades will be discussed in detail in Chapter 4.

3.2 One-Dimensional Theory for Ideal Rotor

A simple 1-D model for an ideal rotor must be examined before use of the blade element momentum (BEM) method. As stated earlier, a wind turbine extracts mechanical energy from the kinetic energy of the wind and the rotor plays a key role in accomplishing this objective. A rotor in a simple 1-D model is a permeable circular disc. The disc is considered an ideal example because it is frictionless and

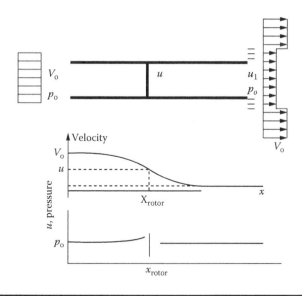

Figure 3.1   Drag dynamics.

has no rotational velocity component in its wake. These are the two fundamental conditions for an ideal rotor.

The disc acts as a drag device and slows the wind speed from the initial wind speed V_0 far upstream of the rotor to wind speed u at the rotor plane and finally to u_1 as illustrated in Figure 3.1.Under these conditions, the streamlines must diverge as shown in the figure. Drag develops from the pressure drop over the rotor. This pressure drop depends on the air density (ρ), initial wind speed (V_0), and the final wind speed at the rotor plane u_1 which will be evident from the expression for the pressure drop across the rotor.

A small pressure rise is expected upstream of the rotor, increasing from pressure p_1 to p before a discontinuous pressure drop (Δ p) over the rotor. Note during the downstream action of the rotor, the pressure returns continuously to the atmospheric level. During this process, the Mach number is very small, the air density is constant, and the axial wind velocity decreases from V_0 to u_1 as illustrated in Figure 3.1. The relationship between the pressure and axial velocity is also illustrated in the figure. Based on the assumptions for an ideal rotor, it is possible to establish mathematical relationships among the various wind velocities involved, the thrust (T) generated, and the absorbed shaft power (P). The expression for the thrust generated can be written as:

$$T = [\Delta p\, A] \tag{3.1}$$

where Δp is the pressure drop across the rotor and A is area of the rotor. The area can be expressed as:

$$A = [\pi R^2] \tag{3.2}$$

where R is the radius of the rotor . Because the flow is stationary, incompressible, and frictionless, no external force acts on the fluid upstream or downstream of the rotor. Under these conditions, the Bernoulli equation is valid; expressions for the pressure upstream and downstream of the rotor can be written as:

$$[p_0 + (1/2)\rho V_0^2] = [p + (1/2)\rho u^2] \tag{3.3a}$$

$$[p - \Delta p + (1/2)\rho u^2] = [p_0 + (1/2)\rho u_1^2] \tag{3.3b}$$

Combining these two equations, the expression for the change in pressure is:

$$\Delta p = [(1/2)\rho \ (V_0^2 - u_1^2)] \tag{3.4}$$

where ρ is the air density (1.225 Kg/m³), p is the pressure rise from the atmospheric pressure level at p_0, V_0 is the wind speed far upstream of the rotor, u is the wind speed at the rotor plane as illustrated in Figure 3.1, and u_1 is the wind speed in the wake.

3.2.1 Axial Momentum Equation in Integral Format

The axial momentum equation in integral format [1] can be applied to the circular control volume with the cross-sectional area of A_{cv} as shown by a dashed line in Figure 3.2 and can be written as:

$$[\partial/\partial t \iiint_{cv} \rho u \ (x,y,z) \ dx \ dy \ dz + \iint_{cv} u(x,y,z)\rho V.dA] = [F_{ext} + F_{pres}] \tag{3.5}$$

where **dA** is a vector pointing in the normal direction of an infinitesimal portion of the control surface with a length equal to the area of this element, **V** is the wind speed vector normal to the area of the element involved, F_{press} is the axial component of the pressure forces acting on the control volume, and F_{ext} is the external force component parallel to the pressure force. Since the flow is assumed to be stationary, the first term in Equation (3.5) is zero and the last term is also zero because the pressure has the same atmospheric magnitude at the end planes and acts on same area of the rotor. Note the force from the pressure has no axial component on the lateral boundary of the control volume (cv) as shown in Figure 3.2. Under the assumptions of an ideal rotor, Equation (3.5) can be rewritten as:

$$[\rho u_1^2 A_1 + \rho V_0^2 \ (A_{cv} - A_1) + (dm_{side}/dtV_0) - \rho V_0^2 A_{cv}] = [- T] \tag{3.6}$$

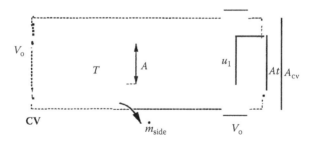

Figure 3.2 Axial momentum.

where T is the thrust acting in the opposite direction generated from the pressure drop over the rotor which reduces the original wind speed V_0 to u_1 as mentioned earlier; (de_{mised}/dt) is the parameter derived from the conservation of the mass equation. The expression for the conservation of the mass can be written as:

$$[\rho A_1 u_1 + \rho V_0 A_{cv} - \rho V_0 A_1 + dm_{side}/dt] = [\rho V_0 A_{cv}] \tag{3.7}$$

This equation provides the expression for (dm_{side}/dt) which can be written as:

$$[dm_{side}/dt] = [\rho A_1 (V_0 - u_1)] \tag{3.8}$$

Using the conservation of mass concept and combining Equations (3.6) and (3.8), the expression for the torque can be written as:

$$T = [\rho u A (V0 - u1)] \tag{3.9a}$$

$$= [(dmside/dt) (V0 - u1)] \tag{3.9b}$$

Using and rearranging Equations (3.1) and (3.9 a):

$$[(1/2) \rho A (V_0^2 - u_1^2)] = [\rho u A (V_0 - u_1)] \tag{3.10}$$

This equation after rearrangement can be rewritten as:

$$[(\rho u A (V_0 - u_1)] = [(1/2) \rho A (V_0 + u_1) (V_0 - u_1)]$$

This means the expression for the wind velocity at the rotor plane can be written as:

$$u = [(1/2) (V_0 + u_1)] \tag{3.11}$$

It is evident from Equation (3.11) that wind velocity in the rotor plane (u) is the mean of the wind speed V_0 and the final speed in the wake (u_1).

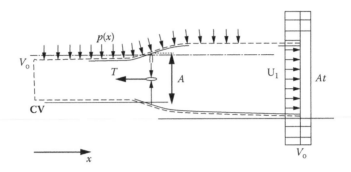

Figure 3.3 Wind turbine with alternate control volume.

3.2.2 One-Dimensional Momentum Theory Using Alternate Control Volume

Wind turbine configuration with an alternate control volume is shown in Figure 3.3. The force generated by the pressure distribution along the lateral walls of the control volume F_{press} is unknown except the force generated by the net pressure contribution F_{press}. In the case of alternate control volume, there is no mass fluid flow through the lateral boundary of the control volume due to its alignment with the streamlines as illustrated in Figure 3.3. The axial momentum equation for the torque (T) in this case can be written as:

$$T = [\rho\, u\, A\, (V_0 - u_1) + F_{press}] \tag{3.12}$$

The force created by the pressure on the control volume following the streamlines is zero as depicted in Figure 3.3. Because of frictionless flow, the wind energy does not change from inlet port to outlet port. The expression for the turbine shaft power P can be written as:

$$P = [energy/time] = [(1/2)\, m\, V^2]/[time] = [watts\ second/second]$$

$$= [watts] \tag{3.13}$$

The shaft power can be obtained by using the integral energy Equation (3.5) on the control volume as shown in Figure 3.3. Inserting the parameters shown in Figure 3.3 into Equation (3.13), the expression for shaft power can be written as:

$$P = (dm/dt)\, (1/2)[V_0^2 + (2P_0/\rho) - u_1^2 - (2P_0/\rho)]$$

$$= [(1/2\ dm/dt)\, (V_0^2 - u_1^2)] \tag{3.14}$$

From Equations (3.9 a) and (3.9 b), the derivative of mass can be written as:

$$[dm/dt] = [\rho \, u \, A] \tag{3.15}$$

Inserting Equation (3.15) into Equation (3.14), the expression for shaft power is reduced to:

$$P = [(1/2) \, \rho \, u \, A] \, [V_0^2 - u_1^2] \tag{3.16}$$

If the induction factor I_f is defined as:

$$I_f = [1 - (u/V_0)] \text{ because } [u = (1 - I_f) \, V_0] \tag{3.17}$$

inserting Equation (3.17) into Equation (3.11) and rearranging the terms yields:

$$u_1 = [2 \, u - V_0] = [(2 \, u/V_0) - 1] \, V_0 \tag{3.18}$$

$$= [2 \, (1 - I_f) - 1] \, V_0$$

$$= [1 - 2 \, I_f] \, V_0 \tag{3.19}$$

Inserting Equations (3.17) and (3.19) into Equation (3.16) allows the expression for shaft power to be written as:

$$P = [(1/2) \, \rho \, u \, A] \, [V_0^2 - (1 - 2 \, I_f)^2 \, V_0^2]$$

$$= [(1/2) \, \rho \, A \, (u/V_0)] \, [1 - (1 - 2 \, I_f)^2] \, [V_0^3]$$

$$= [(1/2) \, \rho \, A \, (1 - I_f)] \, [1 - 1 - 4 \, I_f^2 + 4 \, I_f] \, V_0^3$$

$$= [(1/2) \, \rho \, A \, V_0^3] \, [(1 - I_f) \, (4 \, I_f) \, (1 - I_f)]$$

$$P = [2 \, \rho \, A \, V_0^3] \, [I_f \, (1 - I_f)^2] \tag{3.20}$$

Inserting Equation (3.19) into Equation (3.9 a) yields the expression for torque:

$$T = [(\rho \, u \, A)] \, [V_0 - (1 - 2 \, I_f) \, V_0] = [(\rho \, A) \, (u/V_0) \, (V_0^2) \, (2 \, I_f)]$$

$$= [2 \, \rho \, A \, V_0^2] \, [I_f \, (1 - I_f)] \tag{3.21}$$

where I_f is the axial induction factor and ρ is the density of the rotor material (mass/volume). The available power in a cross-section equal to swept area A by the turbine rotor can be given as:

$$P_{avail} = [(1/2)\ \rho\ A\ V_0^3] \qquad (3.22)$$

3.2.3 Power Coefficient for Ideal One-Dimensional Wind Turbine

The power coefficient (C_p) of the turbine can be written as:

$$Cp = [\text{shaft power/available power}]$$

$$= [\text{Equation } (3.20)/\text{Equation } (3.22)]$$

$$= [2\ \rho\ A\ V0^3]\ [I_f\ (1 - I_f)2]\ /\ [(1/2)\ \rho\ A\ V03]$$

$$= [4\ I_f\ (1 - I_f)2] = [8\ I_f\ (1 - I_f)] \qquad (3.23)$$

The numerical relationships of the various variables are visible in Figure 3.4.

3.2.4 Thrust Coefficient for Ideal One-Dimensional Wind Turbine

The expression for the thrust coefficient for an ideal 1-D wind turbine is:

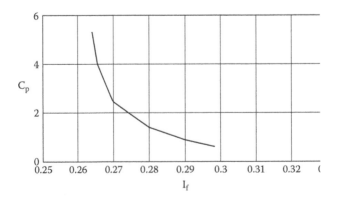

Figure 3.4 Relationship of power coefficient variable for 1-D wind turbine.

$$C_T = \text{[thrust generated by shaft]/[thrust available]}$$

$$= \text{[Equation (3.21)]}/[(1/2)\ \rho\ A\ V_0^2]$$

$$= [2\ \rho\ A\ V_0^2]\ [I_f\ (1 - I_f)]/[\ (1/2)\ \rho\ A\ V_0^2]$$

$$= 4\ I_f\ (1 - I_f)] \tag{3.24}$$

Calculated values of torque coefficients are summarized in Figure 3.5. The maximum power coefficient for the ideal 1-D wind turbine can be obtained by differentiating Equation (3.23) with respect to axial induction factor:

$$[dCp\ /\ dI_f] = 4[(-2I_f)\ (1 - I_f) + (1 - I_f)2] = 4[(1 - I_f)\ (-2I_f + 1 - I_f)]$$

$$= 4[(1 - I_f)\ (1 - 3I_f)] \tag{3.25}$$

Computed values of power coefficients and torque coefficients for an ideal 1-D wind turbine as a function of axial induction factor (I_f) are summarized in Table 3.1. These computed values indicate that the maximum power coefficient occurs when parameter I_f is 0.3, while the maximum value of torque coefficient occurs when the axial induction factor is equal to 0.5. These maximum values of coefficients can be

Table 3.1 Computed Values of Power Coefficient and Torque Coefficient as Function of Axial Induction Factor

Axial Induction Factor (I_f)	Power Coefficient (C_p)	Torque Coefficient (C_T)
0.0	000	000
0.1	0.324	0.360
0.2	0.512	0.640
0.3	0.588	0.840
0.4	0.576	0.960
0.5	0.500	1.000
0.6	0.384	0.960
0.7	0.252	0.840
0.8	0.128	0.640
0.9	0.036	0.360
1.0	000	000

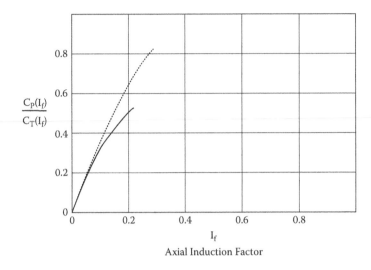

Figure 3.5   Calculated values of torque coefficients.

verified from the curves shown in Figure 3.4. It is important to mention the theoretical maximum power coefficient is equal to 16/27 or 0.5926 when I_f is to equal 0.3 for an ideal wind turbine; this is known as the Betz limit. On the other hand, the theoretical maximum torque coefficient of unity occurs when the axial induction factor equals 0.5 as shown in Figure 3.5 and by the calculations summarized in the figure.

Note that the measured data on thrust coefficients obtained by various aerodynamic scientists [2] indicate slightly higher values as a function of axial induction factor for different rotor states including windmill state, turbulent wake state, and vortex ring state. The measured data further indicate that if the momentum theory were valid for higher values of induction factor, the wind velocity in the wake state would become negative; this can be seen readily in Equation (3.19). In other words, if the axial induction factor exceeds 0.5, the momentum theory is not valid in the turbulent wake state and the wind velocity V_0 would become negative, as shown by Equation (3.19). Furthermore, the expansion of the wake increases with the increase in the thrust coefficient C_T and also the velocity jump from V_0 to u_1 in the wake state.

Based on the forgoing arguments, a high thrust coefficient (C_T) and thus a high axial induction factor I_f present a case for a wind turbine capable of operating at low wind speeds. The reason that the momentum theory is not valid for inductor factor values greater than about 0.4 is that the free shear layer at the edge of the wake becomes unstable when the velocity jump ($V_0 - u_1$) becomes too high and eddy currents are formed (Figure 3.6). It is interesting that the turbulent wake state is produced by unstable shear flow at the wake edge as illustrated by Figure 3.7.

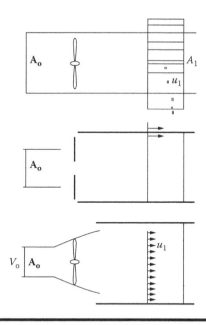

Figure 3.6 Unstable wake caused by velocity jump that generates eddy current.

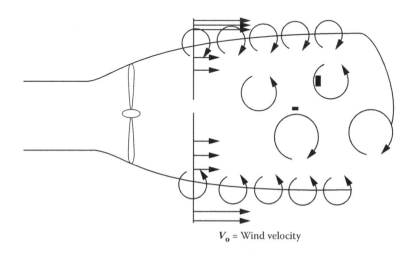

V_o = Wind velocity

Figure 3.7 Turbulent wake state produced by unstable shear flow at wake edge.

3.2.5 Effects of Rotation

The wake of an ideal rotor exhibits no rotation; this means the axial induction factor is zero. Since a modern wind turbine consists of a single rotor without a stator, the wake will possess some rotation that can be verified from the Euler equation for wind turbines. When this equation is applied to an infinitesimal control volume of thickness (dr), the expression for the power contained by the infinitesimal control volume is:

$$dP = [(dm/dt) \, \omega \, r \, C_\theta] = [(2 \, \pi \, r)^2 \, \rho \, u \, \omega \, C_\theta \, dr] \tag{3.26}$$

where m is the mass, ω is the angular velocity or rotational speed, ρ is the air density, u is the axial wind velocity through the rotor, r is the radius, and C_θ is the azimuthal component of the absolute velocity $\mathbf{C} = (C_r, C_\theta, Ca)$, where these absolute velocity components are parallel to the radius, azimuth, and axis, respectively, as illustrated in Figure 3.8. The figure shows the velocity triangle for a section of the rotor and identifies various velocity components downstream of the blade. The blade has a tangential velocity component C_θ in the opposite direction of the blade as shown. The expansion of the wake and the velocity jump for the 1-D ideal wind turbine can also be observed in the figure.

For a given power level P and wind speed, the azimuthal velocity component C_θ in the wake decreases with the increase in the rotational speed ω of the rotor. It is necessary for a wind turbine to have a high rotational speed of the rotor to achieve high efficiency and minimize the kinetic energy loss contained in the rotating wake. Since the axial velocity V_0 through the rotor is defined by Equation (3.17), shown on page 85, in terms of axial induction factor, the rotational speed in the wake can written:

$$C_\theta = [2 \, I_f' \omega \, r] \tag{3.27}$$

$$[V_0 \, (1 - I_f)] \tag{3.28}$$

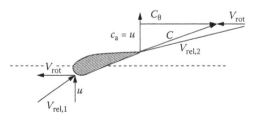

C_0 = Azimuth velocity component in the wake
C_a = Axial velocity of the rotor
C = Absolute velocity of the blade in the down-stream
V_{rot} = Rotor velocity
μ = Axial velocity through rotor

Figure 3.8 Absolute velocity components.

where I_f' indicates the induction factor in the wake. Inserting Equation (3.27) into Equation (3.26) yields:

$$dP = [4 \pi \rho \omega^2 V_0 I_f(1 - I_f) r^3] \, dr \qquad (3.29)$$

The total rotor power can be determined by integrating dP from 0 to R as:

$$P = [4 \pi \rho \omega^3 V_0 \int I_f' (1 - I_f) r^3 \, dr]$$

Local angles of attack below stall conditions based on the potential flow theory in which the reacting force is perpendicular to the local wind velocity as seen by the rotor blade [3].

The power coefficient in a non-dimensional form can be obtained by integrating the following expression with λ ranging from 0.5 to 7.5; it can be written:

$$C_p = [(8/\lambda^2) \int I_f(1 - I_f) x^3 \, dx] \qquad (3.29a)$$

$$\lambda = [\omega R/V_0] \qquad (3.29b)$$

where λ is the tip speed ratio that can vary from 0.5 to 7.5 for modern wind turbines and V_0 is the wind speed. Inserting the values of 0.3 for parameter I_f, 0.5 for parameter I_f' and various values for λ, one can obtain the optimum power coefficient with the Betz limit of 16/27, which is derived for zero rotation in the wake when the parameter I_f' is equal to zero. Computed optimum power coefficient including the

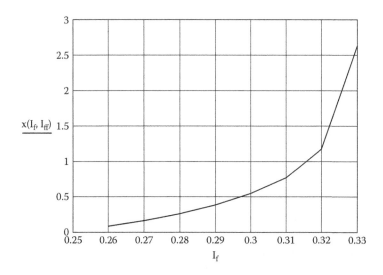

Figure 3.9   Parameter x as function of axial induction factors.

wake rotation effect with the Betz limit will be different from the previous computed value of power coefficient with zero wake effect.

$$x^2 I_f' (1 + I_f') = [I_f (1 - I_f)] \quad\quad\quad (3.30)$$

where

$$x = [\omega \, r/V_0] \quad\quad\quad (3.31)$$

Computed values and a plot of parameter x as a function of axial induction factors are depicted in Figure 3.9. For local angles of attack below rotor stall conditions and to meet the optimization requirements, one can write an expression involving parameters I_f and I_f' as:

$$[(1 - I_f) \, dI_f'/d_f] = I_f' \quad\quad\quad (3.32)$$

Differentiating Equation (3.30) with respect to parameter I_f, one gets:

$$[(1 + 2 I_f') (dI_f'/dI_f) x^2] = [1 - 2 I_f] \quad\quad\quad (3.33)$$

Inserting Equation (3.32) into Equation (3.33) and comparing the coefficients of this new equation with those of Equation (3.30), the result is:

$$[1 + 2 I_f'/1 + I_f'] = [(1 - 2 I_f) / I_f] \quad\quad\quad (3.34)$$

Solving this equation and rearranging the terms yields:

$$I_f' = [(1 - 3 I_f)/(4 I_f - 1)] \quad\quad\quad (3.35)$$

Inserting various values of I_f in Equation (3.35) allowed values of I_f' to be obtained, summarized, and plotted as a function of I_f in Figure 3.8. Inserting these values into Equation (3.30) produced values of x plotted as a function of I_f as illustrated in Figure 3.9. Note in Figures 3.8 and 3.9 that parameter I_{ff} denotes a variable I_f'.

3.2.6 Tip Speed Ratio of Rotor

The tip speed ratio of a wind turbine rotor depends on the number of blades deployed by the rotor. Most commercial wind turbines have three-bladed rotors. There are, however, wind turbines with two blades and also with single blades. The major advantage of fewer blades is the resulting decrease of weight and the costs of the rotor and components. However, maximum capture of wind energy is only possible with a large number of blades. Trade-off analysis of the number of blades and turbine efficiency must be performed to ensure cost-effective performance. For high-capacity wind turbines, the connection of the blades and the hub must be rigid to assure mechanical integrity of the system under a variety of wind conditions. In case of a wind turbine with one or two blades, proper mounting can ensure that they are flexible in the vertical plane. On a so-called teetering hub, two blades can teeter a few degrees across the hub, which reduces the loads on the turbine, thereby yielding higher dynamic stability of the system.

To extract wind energy in an efficient way, the rotor must have an appropriate rotational speed relative to its size, i.e., correct rotor diameter and the wind speed. In other words, the rotor must have an efficient tip speed ratio. The author's studies of tip speed requirements indicate that the tip speed ratio of a wind turbine is strictly dependent on the number of blades deployed by the rotor. The studies further indicate that fewer blades should increase the tip speed ratio. In other words, for wind turbines with the same rotor diameters, one-bladed turbines need higher rotational speeds than two-bladed turbines, which in turn need higher rpm values than three-bladed turbines and also need sophisticated gear box design.

An intimate relationship exists among tip speed ratio (λ), rotational speed (ω), and rotor size (R). The tip speed ratio is defined as the ratio between the speed at the tip of the rotor blade and the undisturbed wind speed (V_0). This means that expression for the tip speed ratio can be written:

$$\lambda = [V_{tip}/V_0] \tag{3.36}$$

It is important to remember that at a given rotational speed the tip speed increases with the length or radius of the blade. Rotational speed (ω) is usually expressed in revolutions per minute (rpm) and tip speed and the wind speed are expressed as meters per second (m/sec). The tip speed can be written as:

$$V_{tip} = [\omega \, 2 \, \pi \, R/60] \text{ m/sec} \tag{3.37}$$

where ω is the rotational speed (rpm) and R is the rotor radius or length (m). Computed values of tip speed as a function of rotor radius and rotational speed are summarized in Table 3.2.

Note that large-capacity wind turbines have longer blades that operate at rotational speeds significantly below 50 rpm to reduce environmental noise level and maintain structural integrity of the turbine. When the blades of a wind turbine rotate, the speed at the tip of the blade is higher than speed in the middle of the blade. For example, for a wind turbine with a 20-m blade length and a rotational speed of 30 rpm, the tip speed will be 60 m/sec but the speed at the middle of the blade will be only 30 m/sec. Figure 3.10 shows the relationships between number of blades and tip speed for wind turbines with six, three and two blades.

Since the speed of the blade segments increases from the root of the blade to the tip of the blade, the apparent wind direction will also change. If one moves

Table 3.2   Tip Speed of Rotor as Function of Radius and Rotational Speed

Rotor Radius, R (m)	Rotational Speed, ω (rpm)				
	10	15	20	25	30
10	10	15	20	25	30
20	20	30	40	50	60
30	30	45	60	75	90
40	40	60	80	100	125
50	50	75	100	125	150

from root to tip of the blade, the apparent wind direction will move toward the vertical plane. Therefore, to achieve the same angle of attack along the entire blade axis, the blade must be twisted as illustrated in Figure 3.10. Apparent wind direction along the rotor blade axis with a wind speed of 9 m/sec is evident due to blade twist. The angle (φ) of the apparent wind direction with respect to the vertical plane varies from 27 degrees at the root of the blade to 6 degrees at the tip of the blade as shown in Figure 3.10. It is important to point out that by twisting the blade so that the angle φ decreases toward the tip, the angle of attack (α) can be kept constant for a given wind condition. The angle of apparent wind direction can be expressed as:

$$\varphi = [\alpha + \beta] \qquad (3.38)$$

where α is the angle of attack (kept as small as possible) and β is the blade angle, normally referred to as pitch angle.

It is important to mention that the undisturbed wind speed (V_{undist}) will decrease to about 2/3 from the assumed wind speed of 9 m/sec just in front of the rotor disc. Estimated values of angle of apparent wind direction with respect to the vertical plane and the magnitudes of undisturbed wind speed at various blade locations are summarized in Table 3.3. These tabulated values provide estimates of undisturbed wind velocities at various locations along the blade radius or length. It is clear that wind speed is at maximum at the tip and minimum at the root. This means that centrifugal forces and the bending moments will be enormous at the tip of a blade.

3.2.6.1 Properties of Airfoils

Efficient design of a wind turbine is strictly dependent on the airfoil configuration and properties. The properties of an airfoil can be obtained from a diagram based on results of the profiles obtained from a wind tunnel. These test results first show

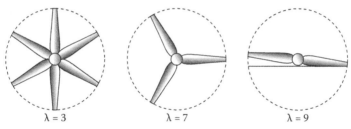

$\lambda = 3$ $\lambda = 7$ $\lambda = 9$

(a) Tip speed ratios for a wind turbine using 6 blades, 3 blades, and 2 blades

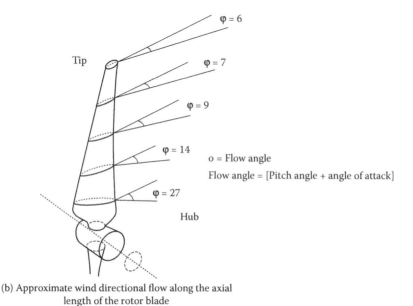

$\varphi = 6$

$\varphi = 7$

Tip

$\varphi = 9$

$\varphi = 14$

o = Flow angle

Flow angle = [Pitch angle + angle of attack]

$\varphi = 27$

Hub

(b) Approximate wind directional flow along the axial length of the rotor blade

Figure 3.10   Relationship of number of blades and tip speed.

the relationship between the angle of attack and the coefficient of lift (C_L), and second, a gliding ratio diagram showing the relationship between lift coefficient and drag coefficient (C_D). If the blade angle and the rotational speed are constant, the angle of attack, lift coefficient, drag coefficient, and gliding ratio will change continuously at different parts of the blade along the blade axis.

It is important to point out that the angle of attack plays a critical role in determining the stall point in an airfoil diagram in which lift coefficient is plotted as a function of angle of attack. An airfoil diagram describes the starting lift with an angle of attack as small as −4 degrees, which reaches a maximum at 15 degrees, then decreases. At such a large angle of attack, the airstream cannot stick to the airfoil surface, thereby creating a turbulence effect under which the

Table 3.3   Undisturbed Wind Speed in Front of Rotor Disc as Function of Blade Radius

Location at Blade Radius	Angle β (degrees)	Undisturbed Wind Speed, Vundist (m/sec)
Blade tip	6	60
0.8 radius	7	48
0.6 radius	9	36
0.4 radius	14	26
0.2 radius	27	12

airfoil begins to stall. Unity list coefficient is possible with an attack angle as high as 7 degrees where an airfoil is most efficient. Some wind turbine designers prefer to use an angle of attack equal to 0.1 radian or 5.73 degrees to achieve efficient and safe turbine performance with no possibility of stall regardless of the wind conditions.

Various airfoil design configurations have been investigated and developed for wind turbine designs, with emphasis on the structural integrity of the rotor blades. When a stream of air passes an airfoil, drag (D) is created in the apparent wind direction and the lift (L) is perpendicular to the drag. These two forces have a resultant force L_{res} Consisting of two components: the useful or circumferential force and the useless force called thrust. The circumferential force is considered useful because, when applied in the plane of rotation, it provides rotational capability to the rotor. Because of rotational capability, creation by an airfoil of a strong circumferential force is the principal requirement of a wind turbine.

Note that a blade profile can have several different thicknesses. The last two digits in the type number of an airfoil indicate relative thickness (i.e., thickness with respect to width) expressed as a percent. For example, for airfoil configuration NACA4412, the maximum thickness of the airfoil is 12% of the width. In case of a wind turbine, the lift from the rotor blades is used to make the rotor revolve, but circumferential force is not the same as lift force. Furthermore, the lift is always applied perpendicular to the apparent wind direction. A rotor blade has a certain angle to the plane of rotation, while the airfoil offers some friction or drag (**D**) that is applied to the apparent wind direction as illustrated in Figure 3.11. These forces are responsible for the development of a circumferential force $\mathbf{F_{CIRC}}$ acting in the plane of rotation. The thrust force perpendicular to the plane of rotation is called $\mathbf{F_{thrust}}$. The magnitude of the circumferential force is small compared to that of thrust, but the power produced is large because the rotational speed is very high as shown in Figure 3.11.

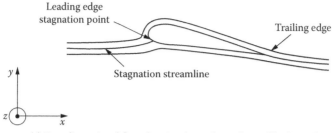

(a) Two-dimensional flow showing the trailing edge and leading edge

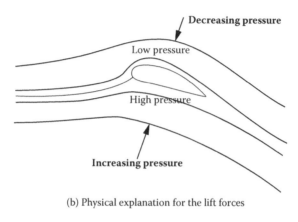

(b) Physical explanation for the lift forces

Figure 3.11   Components of circumferential force.

3.3 Two-Dimensional Aerodynamic Model

Wind turbine rotor blades are long structures. The span-wise velocity component is much lower than the stream-wise velocity component. Under these conditions, it is assumed in many aerodynamic models that the air flow at a given radial position is two-dimensional and that two-dimensional airfoil data can be applied. Note that the two-dimensional flow represents a plane and if this plane is described by the x, y, z coordinate system illustrated in Figure 3.11(A), the flow velocity component in the z direction will be zero.

3.3.1 Airfoil Configuration for Two-Dimensional Aerodynamic Model

It is necessary to extrude an airfoil into a wind of infinite span to realize a 2-D flow. In actual practice, the chord and twist in a real wing section change along the span. The wind starts at the hub and ends at the tip. However, for long slender wings used in wind turbines, aerodynamic scientists have shown that local 2-D data on the forces can be used if the angle of attack is corrected according to the trailing vortices behind the wing. They further claim that the 2-D aerodynamic model is practical for wind turbine rotor design.

The reacting force from the air flow has two components, namely a component perpendicular to the flow velocity and a component parallel to the flow velocity. If an airfoil is designed for an aircraft, the lift-to-drag (L/D) ratio should be maximized. For airfoils designed for applications to wind turbines, the drag-to-lift ratio should be maximized and the drag must be balanced by a propulsion system to maintain a constant speed. The smaller the drag, the smaller will be the required engine. The expressions for the drag (C_D) and lift (C_L) coefficients can be written as:

$$C_D = [D/(1/2) \rho V_0^2 c] \qquad (3.39)$$

$$C_L = [L/(1/2) \rho V_0^2 c] \qquad (3.40)$$

where D is the drag force, L is the lift force, ρ is the air density, c is the chord or length of the airfoil, and V_0 is the flow velocity. The chord is defined as the line from the trailing edge to the nose.

The physical explanation for lift is that the shape of the airfoil forces the streamlines to the curve around the geometry as illustrated in Figure 3.11(B). Basic fluid mechanics theory states that a pressure gradient is necessary to curve streamlines at a radius of curvature r and at a speed of V. The pressure gradient acts like the centripetal force resulting from the circular motion of a particle. Since atmospheric pressure P_0 is present far from the airfoil, pressure below atmospheric must be on the upper side of the airfoil and pressure above atmospheric on the lower side of the airfoil as shown in Figure 3.11(B).

Drag and lift are dependent on the angle of attack. If the lift and drag coefficients applied along the blades are known, it is easy to compute the force distribution on the blades. Global loads such as power output and the root bending moments of a blade can be determined by integrating this distribution along the span. The BEM method may be used to compute the axial and tangential induction factors as well as the loads on the wind turbine. Vortex methods can be used to calculate the induced velocities involved in computations of induction factors.

3.4 Three-Dimensional Aerodynamic Model for Wing of Finite Length

This section describes qualitatively the flow over a 3D wing and how the span-wise lift distribution changes the upstream flow and local angle of attack. Basic vortex theory, as described in various textbooks, plays a key role for this model. Since this theory is not directly used in the BEM method, only its basic concepts will be applied. The material presented here may be abstract for readers with limited knowledge of vortex theory.

3.4.1 Parameters Affected by Streamlines Flowing over Wing

A wing is simply a beam of finite length with airfoils as cross-sections, thereby creating a pressure difference between the lower and upper sides of the wing. The difference in pressure generates lift force. Furthermore, at the wing tips are leakages where air flows around the wing tips from the lower side to upper side of the wing. The streamlines flowing over the wing surface are deflected inward and streamlines flowing under the wing are deflected outward. Therefore at the trailing edge there will be a jump in the tangential velocity due to the leakage at the tips. This velocity jump will create a continuous sheet of stream-wise velocity in the wake behind the wing and this sheet is called the trailing vortices as shown in Figure 3.12.

The classic literature on theoretical aerodynamics reveals that a vortex filament of strength Γ can model the flow past an airfoil for small angles of attack. This is due to the fact that the flow for small angles of attack has zero viscosity and is strictly governed by the linear Laplace equation. For this particular case, lift can be defined using the Kutta-Joukowski equation:

$$L = [\rho \, V_\alpha \, X \, \Gamma] \qquad (3.41)$$

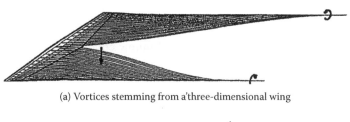

(a) Vortices stemming from a three-dimensional wing

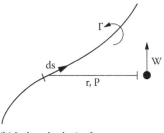

(b) Induced velocity from a vortex
line shown by arrow

Figure 3.12   Trailing vortex.

where α is the air density and V_a is the flow velocity at the angle of attack α. Based on this equation, an airfoil can be thus substituted by one vortex filament of strength Γ and the lift L produced by the wing can be modeled for small angles of attack using a series of vortex filaments oriented in the span-wise direction of the wing [4]. However, a vortex filament cannot terminate in the interior of the fluid but must either terminate on the boundary or be closed. In brief, a complete wind can be modeled using a series of vortex filaments.

The model based on discrete vortices offers reasonable solution. The vortices of the wind known as bound vortices model the lift (L), while the trailing (free) vortices model the vortex sheet stemming from the three-dimensionality of the wind. For a vortex filament of strength Γ, the induced velocity (**w**) at a specific point can be obtained as illustrated in Figure 3.12. Note the relative velocity (V) for a section of a wing is the vector sum of the wind speed and the induced velocity (w). It is important to point out that at the tips of the wing the induced velocity has a value that ensures zero lift.

Based on the above conclusions, it can be stated that for a 3-D wing, lift is reduced compared to a 2-D wing with the same angles of attack and drag forces. Note both effects are due to the downwash induced by the vortex system associated with a 3-D wing configuration. Since the three-dimensionality is limited to the downwash and the span-wise flow is very small compared to the stream-wise velocity, the 2-D data can be used if the geometric angle of attack is modified by the downwash. This assumption is reasonable for long slender wings such as those on a wind turbine. One method to determine the value and the associated induced velocities of the vertices qualitatively is the solution of Prandtl's integral equation. It is important to understand that the vortex system produced by a 3-D wing changes the local inflow conditions. Since the flow is locally 2-D, one cannot apply the geometric angle of attack when estimating the forces on the wind.

3.4.2 Coriolis and Centrifugal Forces

Rotating blade Coriolis and centrifugal forces play an important role in the separated boundary layers that generally occur after a stall condition. In a separated boundary layer, the velocity and the momentum are relatively small compared to the centrifugal force that starts to pump fluid in the span-wise direction toward the tip. When the fluid moves radially toward the tip, the Coriolis force points toward the trailing edge and acts as a favorable pressure gradient. The effect of the centrifugal and Coriolis forces is to alter the 2-D airfoil data after stall. Hansen et al. demonstrated the computed limiting streamline on a modern wind turbine blade moving at a moderately high wind speed. Limited streamlines are flow patterns very close to the surface.

3.4.3 Vortex System behind Modern Wind Turbine

HAWTs are used widely to generate large amounts of electrical energy at minimum cost and complexity. A HAWT rotor is comprised of a number of blades shaped like wings. If a cut is made at a radial distance r from the rotational axis as shown in Figure 3.13(A), a cascade of airfoils can be observed as illustrated in Figure 3.13(B). The local angle of attack α is given by the pitch angle θ of the airfoil; the axial velocity V_a and the rotational velocity V_{rot} are related as:

$$\text{Tan } \varphi = [V_a/V_{rot}] \tag{3.42}$$

where φ is the flow angle and V_a and V_{rot} are the axial and rotational velocities, respectively, at the rotor plane. Since a HAWT consists of rotating blades, a vortex system similar to the linear translating wing must be present. A vortex sheet of the free vortices is oriented in a helical path behind the rotor. The strong vortices are located at the edge of the rotor wake and the root vertices mainly in a linear path along the axis of the rotor as illustrated in Figure 3.14.

The vortex system oriented helically behind the rotor induces an axial velocity component opposite to the direction of the wind and a tangential velocity component opposite to the rotation of the rotor blades. The induced velocity in the axial direction is specified through the axial induction factor a or I_f as aV_0 where V_0 is the undisturbed wind speed. The induced tangential velocity in the rotor wake is defined by a tangential induction factor a′ as (2 a′ ω r). Since the flow does not rotate upstream of the rotor, the tangential induced velocity in the

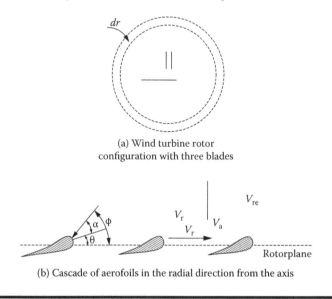

(a) Wind turbine rotor
configuration with three blades

(b) Cascade of aerofoils in the radial direction from the axis

Figure 3.13   HAWT rotor.

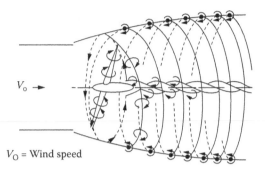

V_O = Wind speed

Figure 3.14   Vortices around HAWT rotor.

rotor plane is thus approximately equal to (a′ ω r), where ω is the angular velocity of the rotor and r is the radial distance from the rotational axis as shown in Figure 3.13. If the induction factors a and a′ are known, a 2-D equivalent angle of attack may be determined.

3.5 Rotor Design Requirements for Wind Farm Applications

Rotor design requirements and configurations vary from application to application. Rotor requirements for wind turbines with lower or moderate capacities not exceeding 10 kW or so are not severe. The number of blades required seldom exceeds two. On the other hand, requirements for two or three blades apply to wind farm applications and tip speeds range from 50 to 70 m/sec when power output exceeds 1 MW.

3.5.1 Rotor Performance

Studies performed by the author on wind turbine rotors indicate that a three-bladed rotor has a tip speed ranging from 45 to70 m/sec to meet high-capacity power levels and yields the best rotor efficiency. The studies further indicate that a two-bladed rotor is 2 to 4% less efficient compared to a three-bladed rotor. A single-bladed rotor with a counterweight for mechanical balance exhibits a further small drop in efficiency that may be about 6% below that of a two-bladed rotor. In brief, a three-bladed rotor is roughly 8 to 10% more efficient than a single-bladed rotor.

As far as the cost and design complexity are concerned, both factors increase with an increase in the number of rotor blades. Weight and procurement costs for three-bladed rotors are higher than those for single or two-bladed rotors. For similarly proportioned blades, rotors with fewer blades must run faster and that can cause problems such as excessive tip noise and erosion. On the other hand, a three-bladed rotor is considered most pleasing esthetically and offers optimum dynamic system stability. Furthermore, the forces acting on the rotor are more evenly balanced and the hub mechanism is simple with a three-blade rotor configuration. One- and two-bladed rotor hubs often present the complications of teetering to limit fluctuating forces as the rotor blades sweep the wind through a varying wind velocity field.

It is important to point out that with a teetered hub, the rotor is hinged at the hub in such a way as to allow the plane of rotation of the rotor to tilt a few degrees backward and forward from the mean plane position. The rocking motion or the teetering of the rotor during each revolution significantly reduces the loads on the rotor blades due to gusts and wind shear and may disturb the dynamic balance of the rotor. A serious dynamic imbalance can damage a wind turbine under severe operating wind environments. In brief, severe dynamic imbalance may pose a serious threat to the safe operation of a wind turbine installation.

3.5.2 Material Requirements for Rotor Blades

The properties of the rotor blade materials depend strictly on the blade dimensions, operating wind conditions, and stresses on the hub structure arising from bending moments. These aspects will be discussed in detail in Chapter 4.

Rotor blades for high-capacity wind turbines are constructed from wood laminates, glass-reinforced plastic (GRP), carbon fiber-reinforced plastic (CFRP), steel, and aluminum. For small wind turbine rotors (diameters smaller than 5 m), the choice of fabrication material is driven by production efficiency and fabrication cost rather than weight, stiffness, and other design requirements. For high-capacity turbine rotors, the blade requirements are more stringent in terms of mechanical integrity and operational reliability. In other words, selection of the rotor blade material depends totally on the mechanical and structural properties of the fabrication materials.

The author will discuss potential ways to use GRP technology to reduce weight and increase the stiffness of the rotor blades in Chapter 4. Some designers have considered high-performance resins such as epoxy in fabrication of rotor blades, but the results are not encouraging. Note that reduction of weight and increase in stiffness are the critical design requirements for longer blades. CFRP blades have been made successfully in prototypes and limited production runs. They have demonstrated the lowest weight and highest stiffness factors but unfortunately the material is costly. It is hoped that the cost of carbon fiber will decrease in the near future as demand increases but the current price of this material is rising steadily. The deployment of this material will not have widespread use for rotor blades until its price drops.

3.5.3 Impacts of Airfoil Characteristics on Rotor Performance

The characteristics of an airfoil significantly impact wind turbine rotor performance. Improvement of rotor performance depends on the following factors:

- High lift-to-drag ratio to improve rotor efficiency over a wide range of wind environments
- Improved stall characteristic
- Low noise production
- Insensitivity to roughness
- Optimum shape for enhanced performance

3.6 Hydrodynamic Analysis of Flow over Rotor

Hydrodynamics flow analysis over a sphere or cylinder is capable of predicting the performance of a wind turbine rotor. This analysis will able to predict the air movement over a solitary spherical hill. In other words, in such an analysis, stream surfaces can be replaced by solid surfaces so that the sphere stream surface will represent the flow over a spherical hill. Similarly, a cylindrical stream flow surface can correspond to flow over a long ridge. It is important to mention that the analysis is not limited to hills shaped as spheres or cylinders; shapes can be approximated by any particular stream surface.

3.6.1 Two-Dimensional Flow Analysis over Sphere

The expressions for the potential (φ) and stream (ψ) functions for the flow over a sphere [5] can be written as:

$$\varphi = [(U \, a^3/2 \, r^2) \cos (\theta) + u \, r \cos (\theta)] \tag{3.43}$$

$$\psi = [- (U \, a^3/U \, r) \sin^2(\theta) + (U \, r^2/2) \sin^2(\theta)] \tag{3.44}$$

where the radial distance r is measured from the origin of the coordinate system that coincides with the center of the sphere, a is the radius of the sphere, U is the flow speed at infinity, and θ is an angle measured from the axis of symmetry and the wind is toward the sphere in the direction when angle θ is equal to 180 degrees.

The equations for the tangential (V_θ) and radial (V_r) components of the velocity can be written:

$$V_\theta = U \sin(\theta)[1 + (1/2) \, (a/r)^2] \tag{3.45}$$

$$V_r = - U \cos(\theta)[1 - (a/r)^2] \tag{3.46}$$

The expression for the surfaces of constant flow speed (q) is:

$$q = [V_\theta^2 + V_r^2]^{0.5} \tag{3.47}$$

The surfaces of the constant flow speed (q), normalized by the flow at infinity (U) can be determined from the solution of the following equation:

$$[x^{-6} + 4 x^{-3}\{1 - 3 \cos^2\theta/1 + 3 \cos^2\theta\}] = [4 (q/U)^2 - 1/(1 + 3 \cos^2\theta)] \tag{3.48}$$

where x equals the ratio of sphere radius to radial distance (a/r). This is a quadratic equation in the cube of x or (a/r) and can be solved using standard methods. Surfaces of the constant dimensionless stream function (ψ/U a^2) and constant speed (q/U) can be plotted. The 2-D flow over a sphere for constant speed is shown in Figure 3.15(a).

3.6.2 Two-Dimensional Flow Analysis over Cylinder

Solutions for flow over a cylinder can be given by the potential (φ) and stream (ψ) functions:

$$\Phi = [U (r + a^2/r) \cos\theta] \tag{3.49a}$$

$$\Psi = [U(r - a^2/r) \sin\theta] \tag{3.49b}$$

The tangential velocity (V_θ) and radial velocity (V_r) components can be written:

$$V_\theta = [U \{1 + (a/r)^2\} \sin\theta] \tag{3.50a}$$

$$V_r = [- U \{ 1 - (a/r)^2\} \cos\theta] \tag{3.50b}$$

where a is the radius of the cylinder and r is the distance as defined earlier. Streamlines defined by the constant value of (ψ/Ua), which is equal to variable, can be calculated directly from the solution of Equation (3.49 b) in the following form:

$$[r/a] = [A/2 \sin\theta] + [1 + (A/2 \sin\theta)^2]^{0.5} \tag{3.51a}$$

The lines of constant speed known as isovelocity contours can be calculated as follows:

$$[(r/a)^{-4} + 2(1 - 2 \cos^2\theta) (r/a)^{-2}] = [(q/U)^2 - 1] \tag{3.51b}$$

Calculated values of the lines of constant velocity or isovelocity contours for 2-D flow over the sphere and cylinder are shown in Figure 3.15(A) and Figure 3.15(B).

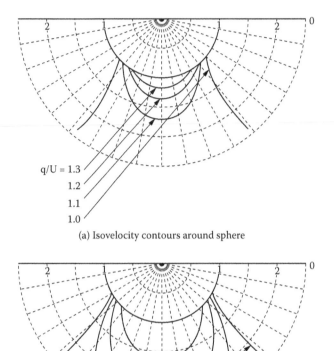

(a) Isovelocity contours around sphere

(b) Isovelocity contours around a cylinder

Figure 3.15 Contours for two-dimensional airflow over sphere and cylinder.

Aerodynamic scientists have presented various concepts for wind turbines and associated rotors, but most fall into two main classes of wind turbines that extract power by employing lift of the surfaces and turbines that strictly depend on the drag of the rotating elements known as rotors or translating elements. In case of rotors, further distinction can be drawn between machines whose axes of rotation are parallel to the incoming wind and those whose axes are normal to the wind direction. Precise classification is further complicated by machines that combine some or all of the basic features. Additional complications are introduced by machines that employ some form of stationary or moving augmenting devices such as inlet ducts, ejector shrouds, or diffuser ducts. Such devices may improve rotor performance, but impose additional weight, size, and cost penalties. Specific details on the penalties of augmentation devices are discussed in Chapter 2.

As mentioned, the rotor is the most critical element of a wind turbine. Rotor design configuration and performance requirements are different for HAWTs and VAWTS. However, regardless of turbine axis configuration, the structural integrity, dynamic stability, weight, size, and efficiency of a rotor must be given serious consideration during design, development, testing, and evaluation of a wind turbine system. Critical aerodynamic aspects of a rotor must be closely examined to ensure cost-effective performance and safe operation under a variety of wind environments. A rotor design must offer continuous operation with no maintenance. It should be noted that any repair or maintenance of a high-capacity wind turbine rotor will be extremely costly and time consuming because such rotors are located about 100 ft above the ground level.

3.6.3 Power Generated by Windstream

The power generating capability of a wind turbine is based on windstream diameter and the undisturbed wind speed in the vicinity of the turbine location. Terrain studies performed by the author for the purpose of augmenting the average wind speed at a turbine installation location indicate that well rounded-hills and ridges are best suited for augmentation. Under certain circumstances, even a sharp peak could be found suitable for such augmentation. The studies further indicate that hills or ridges with abrupt sides are not suitable sites for augmentation. Power generated in the windstream can be determined using the following empirical formula:

$$P \text{ (MW)} = [(\rho \, \pi/8) \, D^2 \, V^3] \tag{3.52}$$

where ρ is the air density (kilograms per cubic meter), D is the windstream diameter (meters), and V is the wind speed in meters per second. Computed values of power generated by the windstream as a function of wind speed are summarized in Table 3.4. It is important to note that the rotor diameter must be less than the windstream diameter to capture maximum wind energy.

3.7 Summary

Performance capabilities and limitations of various types of rotors used by wind turbines are summarized; emphasis is on reliability, dynamic stability, efficiency, and safe operation. Design requirements and performance specifications for Savonius and Darrieus rotors are discussed, along with details concerning power coefficient and cost-effective operation. Analytical capabilities of 1-D, 2-D, and 3-D aerodynamic models for wind turbine rotors are described. The

Table 3.4 Megawatts Generated by Windstream as Function of Diameter and Wind Speed

Wind Speed (mph/m per sec)	Windstream Diameter (ft/m)			
	50/15.24	100/30.48	150/45.72	200/60. 96
20/8.94	0.080	0.319	0.718	1.276
30/13.41	0.270	1.077	2.424	4.307
40/17.88	0.640	2.552	5.744	10.208
50/22.35	1.250	4.984	11.219	19.932

blade element momentum (BEM) theory and its applications to rotor blades are briefly discussed. The 1-D axial momentum theory for an ideal wind turbine is described using the integral energy equation with emphasis on axial components of pressure forces, conservation of mass theory, control volume around a wind turbine, and the forces generated by pressure distribution along the lateral walls of the control volume. Expressions of power coefficient and thrust coefficient for an ideal one-dimensional wind turbine are derived in terms of rotor design parameters.

Effects of rotation on streamlines and flow parameters are identified for modern high-capacity turbine rotors. The importance of Euler equation parameters that affect the axial induction factor and axial rotor velocity is discussed briefly. Factors that affect the tip speed of a rotor are identified. Tip speed calculation as a function of rotor size or diameter and rotational speed is explained. Computed values of undisturbed wind speed as a function of blade length and blade angle are provided for readers who require comprehensive knowledge of wind turbines. Potential characteristics of airfoils best suited to represent the performance capabilities of 1-, 2-, and 3-D wind turbines are summarized.

Airfoil configurations most appropriate for all three types of aerodynamic models are identified. Wind turbine performance parameters affected by streamlines flowing over a rotor are summarized. The effects of Coriolis and centrifugal forces on rotor performance are identified and shaft power developed by a rotor under various wind conditions is explained. The impact of thrust coefficient variation as a function of axial induction factor and the corresponding rotor states is discussed. As thrust coefficient increases, expansion of the wake increases and causes a velocity jump from the incoming undisturbed wind speed to the wind speed past the rotor.

Vortex theory applicable to a 3-D wing of finite length is reviewed with a focus on trailing vortices. The vortex system behind a wind turbine is examined as are limiting streamlines on a modern wind turbine rotor under high wind speed environments. A cascade of airfoils in a vortex system may be observed as function of local angle of attack, airfoil pitch, axial velocity, and rotational velocity at the rotor plane. Rotor performance requirements are defined.

Hydrodynamics analysis of the wind flow over a rotor is performed in terms of wind conditions and rotor design parameters. Wind turbine designers feel that this type of analysis is vital for understanding the theoretical aspects and operating principles of wind turbines. Two-dimensional flow analysis using both potential and stream functions is performed over spherical and cylindrical rotor surfaces to determine isovelocity contours as functions of wind and other variables. These contours are determined from the lines of constant speeds of rotors as the function of an angle measured from the axis of symmetry. Essentially, isovelocity contours are obtained by solving the equations to determine the axial and tangential components of the rotational and axial speeds of a rotor.

References

[1] A.J. Wortman, *Introduction to Wind Turbine Engineering*, Butterworth, Boston, 1983, p. 46.

[2] M.O.L. Hansen, *Aerodynamics of Wind Turbines*, 2nd ed., James & Jane, London, 1992, p. 18.

[3] T. Wizelius, *Development of Wind Turbine Projects*, Earth Scan Publications, Sterling, VA, 2007, p. 8.

[4] S. Mertems, *Wind Engineering in Building Environments*, Multiscience Publications, 1989, Brentwood, U.K.

[5] J.F. Walker and N. Jenkins, *Wind Energy Technology*, 2002, John Wiley & Sons, Chichester.

Chapter 4

Wind Turbine Blade Design Requirements

4.1 Introduction

Understanding the blade theory applicable to a wind turbine blade is vital because the performance of a wind turbine is based on the number of blades assigned to a rotor and the design parameters of the blade elements. The blade theory states that the outer contour of the blade must be follow aerodynamic considerations and the materials used in its fabrication must be capable of providing the needed structural strength and stiffness. This means aerodynamic considerations and properties of the fabrication materials require comprehensive examination during the design and development of the propeller or rotor blades.

This chapter will investigate potential models of actual flow conditions to determine aerodynamic characteristics and predict performance. Aerodynamic analysis of an element of an airfoil will be undertaken to predict performance under a variety of flow environments. The Bernoulli equation for a stream tube ahead of the propeller disc between the plane very far in front of the propeller and a plane immediately in front will be derived in terms of axial interference factor, tangential interference factor, drag and lift forces, undisturbed wind speed, drag-to-lift coefficient ratio, and relative wind speed.

Expressions for the torque coefficient and power coefficient as a function of tip speed ratio, blade loading coefficient, flow angle, and drag-to-lift coefficient ratio will be derived. Situations under which the purely aerodynamic considerations of the blade area distributions are inevitably modified by the structural considerations to achieve safe and optimum blade performance will be explored. Computed values of variations of pitch angle as a function of dimensionless radius, drag-to-lift coeffi-cient, and blade tip speed ratio will be summarized. Beam theory applicable to blade design will be discussed in detail, with emphasis on aerodynamic considerations. Materials and their properties best suited for the fabrication of the rotor blades will be identified along with discussions about strength and stiffness required to ensure high structural integrity under severe wind environments. Equations for the deflections and bending moments at specific points along the blade length will be derived. Important sources of loading on turbine rotor blades resulting from the earth's gravitational field, inertial loading, and aerodynamic loading on the mechanical integrity of a blade section will be described with a focus on tensile stress, compressive stress, and the deflections at the leading and trailing edges.

4.2 Analysis of Performance of Propeller Blades

It is interesting to mention that propeller-based wind turbines span a very wide range of designs from relatively low angular rate Dutch windmills at one end and very high-speed slender propellers at the other end. Since no specialized theories cover the operation of a propeller-based wind turbine, it is necessary to rely on appropriate models of flow situations best suited for the flow characteristics and other parameters that must be analyzed to predict wind turbine performance. Studies performed by the author reveal that the best engineering model is the one that involves the actuator disc analysis of airfoil elements. Furthermore, vortex blade element analysis essentially involves an infinite number of blade models and yields surprisingly good analytical results.

4.2.1 Aerodynamic Performance Analysis of Blade Elements

The flow at an element of width dr located at radius r on a propeller blade rotat-ing at an angular rate of ω is shown in Figure 4.1. Also shown are all the wind component parameters and aerodynamic forces acting on the blade element under consideration. The expression for the Bernoulli equation for a stream tube ahead of the propeller disk element between a plane very far in front of the propeller and plane immediately in front can be written:

$$[(\rho/2) \, V^2 + p_0] = (\rho/2) \, [V^2 (1 - a)^2 + Vrad^2] + p \tag{4.1}$$

where ρ is the air density (1.225 kg/m³), V is the undisturbed wind velocity, V_{rad} is the radial wind velocity, a is the axial interference factor, p is the pressure imme-diately in front, and p_0 is the pressure in front of the propeller. It is important to mention that the radial velocity is very small and the propeller does not induce any tangential velocity component u. Under these conditions, the Bernoulli equation for the planes in front of the propeller and far downstream, using the assumptions

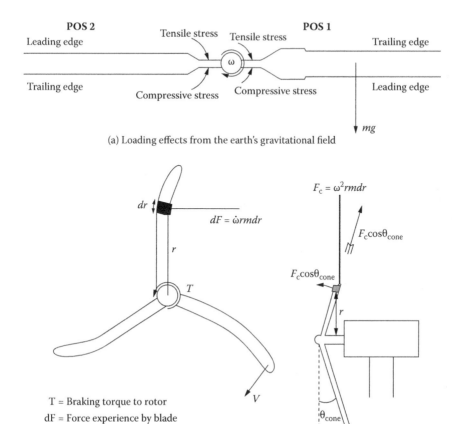

POS 2

Leading edge

Tensile stress

Tensile stress

POS 1

Trailing edge

ω

Trailing edge

Compressive stress

Compressive stress

Leading edge

mg

(a) Loading effects from the earth's gravitational field

dr

$dF = \dot{\omega} rm dr$

r

T

$F_c = \omega^2 rm dr$

$F_c \cos\theta_{cone}$

$F_c \cos\theta_{cone}$

r

V

θ_{cone}

T = Braking torque to rotor

dF = Force experience by blade

r = Radius from the rotating axis

m = Mass per unit length of blade, I = Moment of inertial

(b) Effects of inertial loading on turbine blades

Figure 4.1   Relationships of blade element parameters.

that u is equal to u_1, p_0 is equal to p_1, and each radial velocity is equal to zero, can be written as:

$$[(\rho/2)\{V^2(1-a)^2 + V_{rad}^2 + u^2\} + p + p'] = [(\rho/2)\{(V-V_1)^2 + V_{rad}^2 + u_1^2\} + p_1] \quad (4.2)$$

where

$$p' = [(-\rho V)(V - V_1/2] \quad (4.3)$$

and

$$V_1 = [2 a V] \quad (4.4)$$

Note that parameter u is the tangential velocity component ahead of the propeller.

4.2.2 Thrust and Power on Annular Area of Blade

The expression for the thrust on an annular area (2π rdr) can be written as:

$$dT = [(4 \pi r^3 \rho V^2 a) (1 - a) dr] \tag{4.5}$$

Since the moment imparted to the fluid equals the torque (dT), expression for the power in this infinite annular area can be written as:

$$dP = [(4 \pi r^3 \rho V V_{ang}^2] [a' (1 - a) dr] \tag{4.6}$$

Based on the relationships shown in Figure 4.1, the expression for the elemental torque and power [1] for N rotor blades of chord dimension c is:

$$dT = [(\rho/2) V_{rad}^2 N c (C_L \cos\varphi + C_D \sin\varphi) dr] \tag{4.7}$$

$$dP = [(\rho/2) V_{rad}^2 N c (C_L \sin\varphi - C_D \cos\varphi) r V_{ang} dr] \tag{4.8}$$

When the corresponding terms in Equations (4.5) and (4.7) and Equations (4.6) and (4.8) are equated, the result is:

$$[(a/1 - a)] = [\lambda (1 + \varepsilon \tan\varphi) \cot\varphi \csc\varphi)] \tag{4.9}$$

$$[(a'/1 + a')] = [\lambda (1 - \varepsilon \cot\varphi) \sec\varphi] \tag{4.10}$$

where λ is the blade loading coefficient, ε is the drag-to-lift coefficient ratio, a is the axial interference factor, a' is the tangential interference factor, and φ is the flow angle. The blade loading coefficient is defined as:

$$\lambda = [N c C_L/8 \pi r] \tag{4.11}$$

where N is the number of blades, C is the chord length, r is the radius of the blade, and C_L is the coefficient of lift. The expression for the power coefficient can be written as:

$$C_P = [(dP/dr)/(2 \pi r \rho V^3/2] = [\{4 a(1 - a) X\}(\tan\varphi - \varepsilon/1 + \varepsilon \tan \varphi)] \tag{4.12a}$$

where

$$X = [r V_{ang}/V] \tag{4.12b}$$

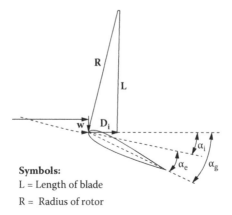

Symbols:
L = Length of blade
R = Radius of rotor

Figure 4.2   Impact of angle of attack and induced drag.

V_{ang} is the angular blade speed and V is the undisturbed wind speed. Figure 4.2 shows the impact of angle of attack on the lift produced and induced drag on a 3-D wing section.

4.2.3 Conditions for Maximum Power Output

Maximum power output can be achieved when the right side of Equation (4.12a) is maximized for a given value of the tangential speed ratio X. Based on the geometrical relationship:

$$[\tan\varphi] = [(1 - a)/(1 + a')]/X \tag{4.13}$$

or

$$[\cot\varphi] = X[(1 + a')/(1 - a)] \tag{4.14}$$

By eliminating a′ using Equations (4.9) and (4.10), Equation (4.13) can be rewritten as:

$$[X \tan\varphi] = [1 - (a \sec^2\varphi/1 + \varepsilon \tan\varphi)] \tag{4.15}$$

Elimination of parameter a from Equation (4.15) leads to:

$$[\tan\varphi] = [\sin\varphi - (X'- \varepsilon) C_{DL}']/[(X' - \varepsilon) \sin\varphi + C_{BL}'] \tag{4.16a}$$

where

$$X' = [X (1 + \varepsilon \tan\varphi)] \text{ and } C_{BL}' = [C_{BL} (1 + \varepsilon \tan)] \tag{4.16b}$$

4.3 Blade Performance

Optimum blade performance for a given drag-to-lift ratio is critical, particularly when calculated as a preliminary estimate. Blade performance is measured in terms of power coefficient C_p, which is a function of the tangential speed ratio X for a given range of values of drag-to-lift coefficient ratio ε.

4.3.1 Power Coefficient

Power coefficient defines the performance of a wind turbine in terms of relative power level as a function of tangential speed ratio and drag-to-lift coefficient ratio. Computed values of the power coefficient as function of tangential speed ratio (X) for various values of ε are summarized in Table 4.1. The calculated values indicate that the maximum values of power coefficients occur for the tangential speed ratio (X) when it varies between 1 and 4 regardless of the values of ε. However, maximum values of power coefficients occur when ε is equal to zero, which may be impossible under certain wind conditions.

4.3.2 Axial Interference Factor

Axial interference factor a plays a key role in derivation of other parameters that are vital for the design and development of a turbine blade. Any variation in axial interference may affect the mechanical integrity of a blade. Studies of aerodynamic scientists indicate that the axial interference factor or coefficient depends on the drag-to-lift coefficient ratio (ε) and tangential speed ratio (X) values. Therefore, any variation of either parameter will affect the variation in the axial interference factor or coefficient a. Calculated values of variation in the axial interference factor as a

Table 4.1   Power Coefficient Calculation as Function of Tangential Speed Ratio (X) and Drag-to-Lift Coefficient Ratio (ε)

Tangential Speed Ratio (X)	Drag-to-Lift Coefficient Ratio (ε)			
	0.01	0.03	0.05	0.075
1	0.51	0.47	0.46	0.43
2	0.55	0.51	0.47	0.42
4	0.56	0.48	0.41	0.32
6	0.53	0.43	0.33	0.22
8	0.52	0.38	0.25	0.12
10	0.50	0.33	0.21	0.05

Table 4.2   Computed Values of Variations in Axial Interference Coefficient (a)

Tangential Speed Ratio (X)	Drag-to-Lift Coefficient (ε)			
	0.01	*0.03*	*0.05*	*0.10*
1	0.328	0.317	0.312	0.294
2	0.328	0.322	0.311	0.285
3	0.335	0.322	0.303	0.274
4	0.334	0.318	0.296	0.242
5	0.334	0.304	0.285	0.212
6	0.333	0.297	0.274	0.175
7	0.331	0.290	0.257	–
8	0.325	0.284	0.242	–
9	0.320	0.280	0.234	–
10	0.318	0.270	–	–

function of drag-to-lift coefficient and tangential speed ratio are summarized in Table 4.2. Based on these values, the variation in the axial interference coefficient is extremely small when the tangential speed ratio varies from 1 to 10 at a drag-to-lift coefficient ε equal to 0.01. The variations in the axial interference coefficient increase as a function of tangential speed ratio, regardless of the values of the drag-to-lift coefficient. This conclusion also applies to power coefficient variations.

4.3.3 Torque Grading Coefficient

Wind turbine output is a function of torque generated by a turbine rotor. The rate of torque generated is strictly dependent on the torque grading coefficient, which is a function of tangential speed ratio (X) and drag-to-lift coefficient (ε). Calculated values of torque grading coefficient variations as a function of these two parameters are summarized in Table 4.3.

It is evident from these calculated values that large variations in the torque grading coefficient occur over a wide range of tangential speed ratios for a given drag-to-lift coefficient. However, moderate variations in the torque grading coefficient are seen at a given tangential speed ratio when the drag-to-lift coefficient varies from 0 to 0.1. Note very large variations in torque grading coefficient occur when the tangential speed ratio exceeds about 8, irrespective of the magnitude of drag-to-lift coefficient. As mentioned previously, both the blade loading coefficient

Table 4.3   Variations of Torque Grading Coefficient as Function of Tangential Speed and Drag-to-Lift Coefficient

Tangential Speed Ratio (X)	Drag-to-Lift Coefficient (ε)			
	0	0.03	0.05	0.10
2	0.295	0.256	0.238	0.205
3	0.196	0.169	0.165	0.158
4	0.165	0.128	0.103	0.062
5	0.128	0.085	0.072	0.034
6	0.092	0.072	0.056	0.023
7	0.082	0.057	0.041	–
8	0.074	0.048	0.033	–
9	0.065	0.039	0.026	–
10	0.059	0.034	–	–

and the relative flow angle are insensitive to large variations of the drag-to-lift coefficient, observed particularly at lower tangential speed ratio (X).

4.3.4 Blade Loading Coefficient

The author's preliminary aerodynamic studies of rotor blades indicate that both the blade loading coefficient λ and the relative flow angle are insensitive to large variations in the drag-to-lift coefficient, particularly at lower values of the tangential speed ratio (X) as illustrated by Table 4.4. The blade loading coefficient is the most critical parameter because its impacts the mechanical integrity of a blade under severe mechanical environments. The blade loading coefficient as defined by Equation (4.11) can be rewritten:

$$\lambda = S \, [C_L/4] \qquad (4.17)$$

where S is the solidity parameter and:

$$S = [N \, c/2 \, \pi \, r] \qquad (4.18)$$

The theory of wind turbines reveals that solidity can be approximated by seven times the blade loading coefficient (λ) during the preliminary design analysis. For a wind turbine with a drag-to-lift coefficient of zero and a tip speed ratio of 10, this corresponds to a solidity of about 1.5% near the blade tip. Blade sections with

Table 4.4 Variations of Blade Loading Coefficient as Function of X and ε Parameters

Tangential Speed Ratio (X)	Drag-to-Lift Coefficient		
	0	0.05	0.10
1	0.125	0.125	0.125
2	0.046	0.045	0.044
3	0.025	0.023	0.020
4	0.015	0.014	0.011
5	0.009	0.008	0.007
6	0.006	0.005	0.004
7	0.0046	0.0038	–
8	0.0035	0.0027	–
9	0.0028	0.0022	–
10	0.0023	–	–

drag-to-lift coefficients of 0.1 and tip speed ratios of 2 will be very inefficient and the corresponding solidities will be close to 45%. Turbine blades with high solidities must be discarded if efficient and stable operation is the design goal. The various parameters were defined under Equation (4.11).The aerodynamic theory of a blade reveals that the distribution of the blade loading coefficient (λ) as a function of dimensionless radius and coefficient of load can be achieved. The theory further reveals that for representative airfoil sections, minimum value of drag-to-lift coefficient (ε) occurs when the coefficient of lift (C_L) varies from approximately 0.5 to 0.8. Under these parametric conditions, the solidity parameter (S) can be approximated by 7λ.

Computed values of variation of the blade loading coefficient as a function of tangential speed ratio (X) and drag-to-lift coefficient (ε) are summarized in Table 4.4. Various types of loadings involved in wind turbine design, namely Earth's gravitational loading, internal loading, and blade loading are shown in Figure 4.1. Various forces and torques generated by these loading effects are also illustrated. It must be stressed that that the purely aerodynamic design considerations for blade area distributions may be modified by structural considerations that may introduce centrifugal loads, aero-elastic effects, and fatigue conditions.

These computed values indicate little change in blade loading coefficient over a tangential speed ratio ranging from 1 to 3, regardless of the magnitude of the drag-to-lift ratio (λ). It is obvious from these tabulated data that only moderate variations occur in the blade loading coefficient when the tangential speed ratio varies from 4 to 10, irrespective value of the drag-to-lift coefficient.

4.3.5 Variations of Flow Angle as Function of Tangential Speed Ratio and Flow Angle

Analytical studies by the author revealed that minute variations in flow angle occur as a function of tangential speed ratio and drag-to-lift coefficient; this may be verified from the computed data summarized in Table 4.5. This means that flow is least affected from the variations in the magnitudes of tangential speed ratio and the drag-to-lilt coefficient. Variations in wind speed as a function of flow angle along the axial length of the rotor blade are shown in Figure 4.3. Note that the flow angle is the sum of the blade pitch angle (β) and the angle of attack (α).

These tabulated data indicate that the variation of the relative flow angles decreases with the increase in the tangential speed ratio regardless of the value of drag-to-lift coefficient. However, the difference between the relative variations is small at a given drag-to-lift coefficient and given tangential speed ratio.

Table 4.5 Variation of Relative Flow Angle (φ) as Function of Tangential Speed Ratio and Drag-to-Lift Coefficient

Tangential Speed Ratio (X)	Drag-to-Lift Coefficient (ε)		
	0	0.05	0.10
1	32	33	34
2	17	18	19
3	13	14	15
4	9	10	11
5	7	8	9
6	6	7	8
7	5	6	–
8	4	5	–
9	3	4	–
10	4	–	–

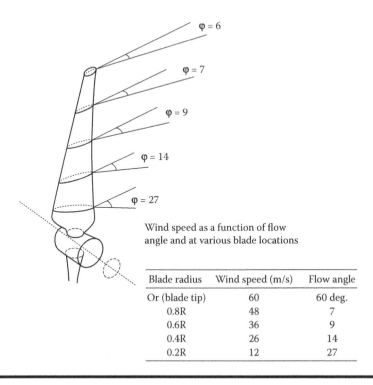

Wind speed as a function of flow
angle and at various blade locations

Blade radius	Wind speed (m/s)	Flow angle
Or (blade tip)	60	60 deg.
0.8R	48	7
0.6R	36	9
0.4R	26	14
0.2R	12	27

Figure 4.3   Impacts of variations in wind speed.

4.3.6 Impact of Tip Speed Ratio and Drag-to-Lift Coefficient on Wind Turbine Power Coefficient

Wind turbine efficiency, also known as power coefficient, is a function of tip speed ratio and drag-to-lift coefficient. The speed ratio and drag-to-lift coefficient are the most critical parameters affecting wind turbine power. The impact of tip speed ratio and the drag-to-lift coefficient on the power coefficient is evident from the computed values in Table 4.6.

Based on these tabulated data, the maximum values of the power coefficient occur when the tip speed ratio varies from 2 to 6 approximately, regardless of the value of the drag-to-lift coefficient. The power coefficient continues to maintain optimum values over a tip speed ratio ranging from 1 to 10, irrespective of the magnitude of drag-to-lift coefficient. If the plots for power coefficient versus tip speed ratio (X_0) are drawn for various values of ε, then the peaks of the curves will be found optimum for the tip speed ratios for wind turbines whose blades have a pitch distribution along the radius that corresponds to the optimum turbine operation for a particular tangential speed ratio and the drag-to-lift coefficient.

It is important to mention that the maxima of the overall power coefficients occur over tip speed ratios ranging from 2 to 5. The variation of the power coefficient

Table 4.6   Impacts of Tip Speed Ratio and Drag-to-Lift Coefficient on Wind Turbine Power Coefficient (C_P)

Tip Speed Ratio (X_0)	Drag-to-Lift Coefficient (ε)			
	0	0.01	0.03	0.05
1	0.430	0.422	0.417	0.402
2	0.513	0.502	0.477	0.452
3	0.555	0.532	0.491	0.448
4	0.562	0.545	0.492	0.449
5	0.550	0.552	0.486	0.428
6	0.578	0.550	0.464	0.402
7	0.580	0.537	0.450	0.386
8	0.585	0.535	0.445	0.358
9	0.586	0.531	0.428	0.333
10	0.587	0.528	0.412	–

with tip speed ratio increases rapidly with increasing magnitudes of the drag-to-lift coefficient and this indicates the high sensitivity of inefficient rotor blades to off-design conditions. If the distribution of the turbine power output is plotted as a function, one will notice that about 5% of the total wind power is produced in the disc area corresponding to 20 to 25% of blade radius.

4.3.7 Variations in Pitch Angle as Functions of Radius

The pitch angle of a blade plays a key role in maintaining a near-uniform rotor speed under variable wind speed environments—a requirement to achieve optimum power output from a wind turbine. Computed values of variations in pitch angles at a given angle of attack of 0.1 radian (5.73 degrees) and as a function of dimensionless radius r/R are summarized in Table 4.7.

These computations assume a minimum drag-to-lift coefficient of zero, which means that ε is equal to ε_0. It is evident from these results that the variations in pitch angle decrease rapidly as a function of tip speed ratio, regardless of the value of the drag-to-lift coefficient. However, moderate variations in pitch angle occur at a given tip speed ratio when the drag-to-lift coefficient varies from 1 to 10. The value shown under the 0.01/5 column means that the variation in the pitch angle is optimum when the drag-to-lift coefficient and the tip speed ratio are 0.01 and 5,

Table 4.7 Degrees of Variation in Pitch Angle as Function of Dimensionless Radius, Drag-to-Lift Ratio, and Tip Speed Ratio

	Drag-to-Lift Ratio/Tip Speed Ratio (ε_0/X_0)		
Dimensionless Radius (r/R)	*0.01/5*	*0.03/3.5*	*0.05/3*
1	35	39	43
3	17	24	26
4	12	19	22
5	8	14	17
6	7	12	14
7	5	9	12
8	4	8	10
9	3	6	8
10	2	5	7

respectively. Similarly, the value shown under the 0.05/3 column indicates that the variation in the pitch angle is optimum when the drag-to-lift coefficient and the tip speed ratio are 0.05 and 3, respectively. As stated earlier, these computations were carried out for an angle of attack of 5.73 degrees—considered ideal for blades used in wind turbines.

All these numerical data are vital aspects of the design and development of blades for wind turbine applications. It is important to point out that blades must be designed for optimum power generating performance and improved mechanical integrity under severe wind conditions. Optimum performance requires a variation of the airfoil sections along the radius to ensure that each element operates at its minimum drag-to-lift coefficient—the principal design consideration. For any value of a drag-to-lift coefficient (ε), the performance of a complete blade corresponds to an interval of one of the curves as illustrated in Figure 4.2. Note this interval starts at a very low value of the tangential speed ratio (X) corresponding to the hub of the blade and terminates when X equals the tip speed ratio (X_0) of the wind turbine and is often used to characterize the particular design. The structural considerations at the hub constitute a different design problem and no attempt is made to assess or achieve aerodynamic performance there. It is not uncommon to use different airfoil sections in the inner and outer regions of the blade, but such fine points of design optimization will not be considered here. The author's studies of blade performance indicate that the effects of tip losses and hub blockage barely degrade wind turbine performance—no more than 3 to 5% and these effects will not be discussed in detail.

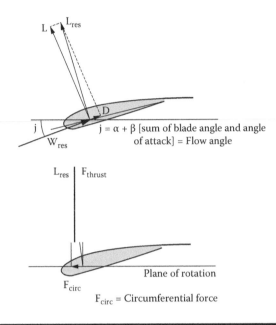

Figure 4.4   Forces exerted on wind turbine blades.

4.3.8 Forces Acting on Blades

A wind turbine blade is subjected to several forces that depend on pitch angle, angle of attack, and geometrical parameters of the blade. These forces include lift, drag, residual lift, forward thrust, and circumferential force as shown in Figure 4.4. The magnitudes of these forces depend on the plane of rotation and flow angle. Note the lift and the circumferential force components acting on the blade play an important role in the performance of a wind turbine rotor. The roles of these forces and aspects of stiffness and mechanical integrity will be discussed.

A section of the blade illustrated in Figure 4.5 can be modeled as a beam. When the stiffness factors EI and GI at different span levels are calculated as a function of geometric parameters and fabrication material properties, the simple beam theory can be applied to compute the stresses and deflections at various locations on the blade. Parameter E is the modulus of elasticity (Young's modulus); G is the modulus of shear elasticity; and I represents moments of inertia at various locations along the blade as shown in Figure 4.5.Specific details on simple beam theory are

Figure 4.5   Calculation of blade stresses and deflections.

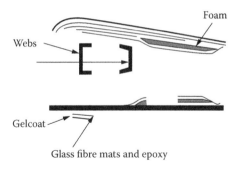

Figure 4.6 Blade fabrication details.

readily available in basic books on mechanics of materials such as Timoshenko's 1972 text.

Various materials used in the fabrication of a blade are described in Figure 4.6. The most common materials used include foam, Gelcoat, glass fibers, and epoxies that have unique characteristics. Fiberglass offers high stiffness, flexural strength, and shear resistance under extreme thermal and mechanical environments. If epoxy materials are used, it is vital to select an optimum housing metal (steel or aluminum) and evaluate the critical performance characteristics of the embedding resins such as viscosity, exothermic properties, and other factors required to ensure blade integrity under severe thermal and mechanical environments.

4.3.9 Mechanical Integrity

Structural reliability and mechanical integrity must be given serious consideration during the design phase. The tilt of the nacelle, the aerodynamic loading as a function of wind speed conditions, the wind speed at the hub of the blade, the flap-wise and edge-wise bending moments of the blade, and the gravitational loading on the blade all affect the mechanical integrity and consequently the reliability of rotor blades. The effect of gravitational loading is very severe on the edge-wise bending moment as a dominant sinusoidal vibration, upon which is superimposed some small high-frequency signals originating from atmospheric turbulence conditions. The flap-wise bending is mostly produced by aerodynamic loads that vary with the turbulent wind fields and is relatively less serious than adverse effects produced by the earth's gravitational field and inertial loading as illustrated in Figure 4.3.

4.3.9.1 Structural Parameters and Stresses

Figure 4.7 illustrates the main structural parameters and various stresses or forces acting on a section of a beam. The structural parameters are described as follows:

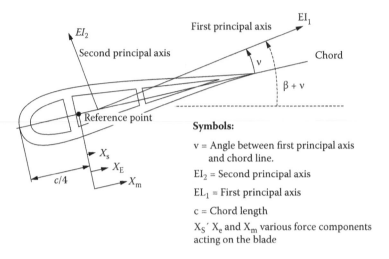

Figure 4.7   Structural parameters and stresses.

EI_1 = bending stiffness about the first principal axis
EI_2 = bending stiffness about second principal axis
GI_v = torsional stiffness
X_E = distance of the point of elasticity from the reference point shown in Figure 4.6
X_m = distance of the center of the blade mass from the reference point
X_s = distance of the shear center from the reference point
β = angle of twist of airfoil section measured with respect to the reference line
v = angle between chord line and first principal axis

The point of elasticity is defined as the point where the normal plane will not give rise to a bending moment of the beam. The s point where the shear stress is concentrated will not give rise to a deflection under shear stress and will not rotate the airfoil. The moment of inertia (I) about one of the principal axes plays a key role in the blade deflection. The impacts of critical quantities such longitudinal stiffness (EA) and moment of centrifugal stiffness (ED_{XYR}) defined in terms of reference coordinate system on the blade must be described in detail. The reference coordinate system (X_RY_R) is illustrated in Figure 4.7.

EA = longitudinal stiffness
ES_{XR} = moment of stiffness about the axis X_R
(X_R, Y_R) = reference coordinate system as indicated by black dot
EI_{XR} = moment of stiffness about axis X_R (Figure 4.8)
EI_{YR} moment of stiffness about axis Y_R (Figure 4.8)
ED_{XYR} = moment of centrifugal stiffness

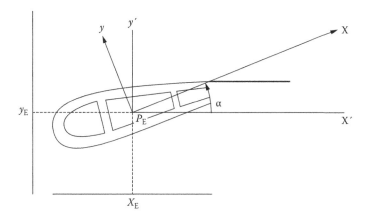

Figure 4.8   Forces acting on wind turbine blades.

Using these quantities, the point of elasticity P_E (X_E, Y_E) can be calculated in the reference system as:

$$X_E = [ES_{YR}] \tag{4.19}$$

$$Y_E = [ES_{XR}] \tag{4.20}$$

Using E (Young's modulus) and ρ (material density), the point (X_E, Y_E) represents the center of mass for the section illustrated in Figure 4.8. If the moment of stiffness inertia and moments of centrifugal stiffness are moved to the coordinate system designated as (X′,Y′), which is parallel to the reference system (X_R, Y_R) as shown in Figure 4.8, the expression for the point of elasticity can be easily written. The angle of attack (α) between the X′ axis and first principal axis and the bending stiffness about the principal axes can be calculated from these equations. Also the stress equation can be written as:

$$\Sigma (x,y) = [E (x,y)][\varepsilon (x,y)] \tag{4.21}$$

where the expression for the strain can be written as:

$$\Sigma (x,y) = M_1 \, y/[EI_1] - M_2 \, x/[EI_2] + N/[EA] \tag{4.22}$$

where the constant σ, ε, and N are positive for tension and negative for compression and the normal Force N can computed from the loadings on the blade. These parameters now can determine the main structural characteristics. Since a wind turbine blade is generally very stiff in torsion, the torsional deflection is normally neglected. Shear center and the torsional rigidity can be calculated using computer analysis.

Table 4.8   Structural Parameters of 2 MW Tjaereborg High-Performance Wind Turbine

Radius (m	EA (GN)	EI_1 (MNm)²	EI_2 (MNm)²	GI_y (MNm²⁾	Mass (Kg/m)	X_E (mm)	X_m (mm)	X_s (mm)	v (deg)	β+V (gdeg)
1.8	36.0	12000	12000	7500	1700	00	00	00	00	00
3.0	6.14	1630	1725	362	330	2	2	0	5.4	14.4
4.5	5.82	1080	1940	328	389	54	159	11	0.94	9.44
6.0	5.10	623	1490	207	347	59	165	13	1.3	9.30
9.0	4.06	2.55	905	92.8	283	63	170	18	1.09	8.09
12.0	3.33	129	557	47.7	235	58	158	15	0.86	6.86
15.0	2.76	64.8	349	24.7	196	51	137	15	086	5.86
18.0	2.33	32.4	221	12.9	166	45	121	16	0.91	4.91
21.0	1.83	15.2	131	6.23	172	41	110	17	0.83	3.83
24.0	1.21	6.04	65.7	2.57	90.3	40	102	16	0.63	2.63
27.0	0.63	1.82	28.1	0.84	52.6	47	108	14	0.16	1.16
30.0	0.21	0.32	9.5	0.18	24.2	82	136	10	–0.52	–0.52

Various structural data for a 30-m turbine blade used on a 2-MW Tjaereborg wind turbine are summarized in Table 4.8. Deflections, various forces acting on the blade, and bending moments of the blade are shown in Figure 4.9. Note the coordinate system used in Figure 4.9 is different from the system shown in Figure 4.8.

4.3.9.2 Shearing Forces and Bending Moments in Presence of External Forces

The magnitudes of shearing forces (Tz, and T_y) and bending moments (M_y and M_z) acting on a blade in the presence of external forces are different from those in the absence of the external forces (p_y and p_z). See Figure 4.10. The shearing forces and the bending moments can be calculated using the following equations [3]:

$$[dT_z/dx] = [- p_z(x)] + [m(x) \, d^2U(x)/dz^2] \tag{4.23}$$

$$[dT_y/dx] = [- p_y(x)] + [m(x) \, d^2U(x) / d_y^2] \tag{4.24}$$

$$[dM_y/dx] = T_z \tag{4.25}$$

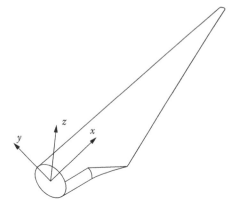

(a) Locations of various axes of the blade

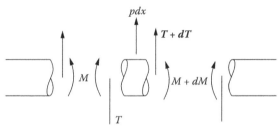

(b) Various bending moments and forces acting on the blade

Figure 4.9   Deflections and bending moments.

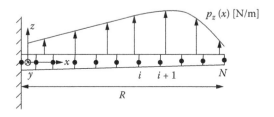

Figure 4.10   Shearing forces and bending moments.

$$[dM_z/dx] = T_y \tag{4.26}$$

where m is the mass and d^2U_z/dz^2 and d^2Uy/dy^2 are the accelerations along the z and y directions, respectively, and U(x) is the velocity in the x direction as illustrated in Figure 4.10. If the blade is at equilibrium, the final (inertial) terms on the right sides of Equations (4.23) and (4.24) approach zero. Although these equations have been taken from Walker [4], the expressions can be derived using the Newton's second law on an infinitesimal section of beam and using the parameters illustrated in Figure 4.9(B). Newton's second law is best suited to solve fluid dynamics problems.

4.4 Application of Beam Theory to Various Turbine Blade Configurations

Aerodynamic studies performed by the author indicate that beam theory can be applied successfully to design various blade configurations. This section describes how a rotor blade whose outer contour adheres strictly to aerodynamic considerations is fabricated to be sufficiently strong and stiff to withstand various forces under unusual wind conditions. Materials used in the past such as wood, aluminum, glass fiber-reinforced plastic (GFRP), and carbon-fiber-reinforced plastic (CFRP) have not demonstrated the required strength and stiffness under extreme wind environments.

4.5 Material Requirements for Blades

It is extremely important to mention that the selection of the blade material depends on several parameters including mechanical strength, stiffness factor, weight and price per unit of length, and fatigue properties of the material. According to turbine blade designers, the fatigue properties of the material are vital. Several blade materials now available are capable of meeting specific structural and aerodynamic requirements. Specific details on their mechanical and aerodynamic specifications are contained in the latest hand books on material science.

4.6 Critical Features of Blade Section

Specific features of an actual turbine blade section can be seen in Figure 4.4. A turbine blade can be modeled as a simple 3-D beam when the stiffness factors EI and GI as illustrated in Figure 4.5 at different span-wise stations are computed. Simple beam theory can be applied to compute various stresses and deflections as shown in the figure. Note that E indicates the modulus of elasticity of the material, G represents the modulus of elasticity for shear, and I is the moment of inertia for different sections of the blade.

A computer model is available to compute structural parameters of a turbine blade with emphasis on rigidity, tensile strength, deflections and bending moments as functions of the materials used in fabricating the blade elements, blade geometrical parameters, airfoil configuration, and angle of attack. Structural parameters must be determined for blade materials considered for turbines operating in harsh mechanical environments to avoid catastrophic damage to tower and the turbine components. It is absolutely essential to compute the deflections of the blade for a given load condition or as an input to a dynamic simulation using aeroelastic codes recommended by blade designers as best suited to provide high mechanical integrity in adverse wind conditions.

4.6.1 Impacts of Bending Moments and Blade Instability on Performance

It is interesting to mention that any rotor blade movement or instability will try to tilt the nacelle over the tower. The yaw motor movement tries to turn the nacelle back on the tower. Sometimes under extreme wind conditions, two bending moments at the root of the tower will tend to disturb mechanical stability even at a moderate wind speed of 11 m/sec. In addition, the flap-wise and edge-wise bending moments, tilt motor and yaw motor bending moments, and atmospheric turbulence may affect turbine blade stability. Flap-wise bending is somewhat reduced by the aerodynamic loads that vary with the turbulence wind field.

4.6.2 Role of Wind Triangle

A wind triangle on a turbine blade element plays key role in predicting the preliminary performance of wind turbines with output capacities not exceeding 10 kW. A device such as the triangle installed on a turbine blade element as illustrated in Figure 4.11 can provide a preliminary indication of blade performance as a function of flow angle (φ), wind speed (V), axial interference factor (a), radius (r), angle of attack, and axial interference factor for turbines with output power levels not exceeding 10 kW. Because computation of structural parameters similar to those summarized in Table 4.8 will not be cost-effective because of 10 × 10 matrix elements, the electrical performance of small wind turbines with optimum power output can be determined by the following equations so long as the angle of attack is below stall level and parameters a and a′ are not independent. If the total induced velocity is in the same direction as the force and perpendicular to the local velocity, the following performance expression for a small wind turbine can be written:

$$[x^2 a' (1 + a')] = [a'(1 - a] \tag{4.25}$$

where

$$x = [w r V_0] \tag{4.26}$$

$$a' = [(1 - 3a)/(4a - 1)] \tag{4.27}$$

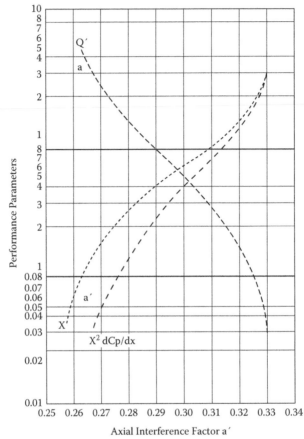

(a) Performance parameters for an ideal windmill

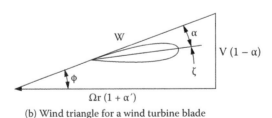

(b) Wind triangle for a wind turbine blade

Figure 4.11   Wind triangle on turbine blade.

where w represents rotational speed. The optimum value for the axial interference factor tends to be 0.333 which is consistent with the optimum power coefficient. Computed values of performance parameters (a, a′, and x) and wind triangle configuration for a wind turbine are shown in Figure 4.11. The figure illustrates the absolute velocity of the blade and hence is referred as the velocity triangle downstream

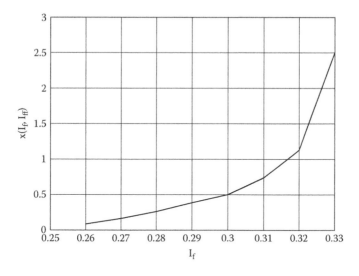

Figure 4.12 Performance curve as function of axial interference.

of the blade. For a given power level P and wind speed, the azimuthal velocity component in the wake decreases with the increasing rotational speed w of the rotor. From an efficiency view, it is highly desirable for a wind turbine to have a high rotational speed to minimize the kinetic energy loss in the rotating wake.

Note that the axial velocity through the rotor is expressed by the axial induction factor or axial interference factor (a). The rotation speed of the blade in wake is given as (2 a′ w r) where a′ is the axial induction factor in the wake, w is the rotation speed of the blade in the wake, and r is the blade radius [5]. The performance curve of a turbine is expressed in terms of parameter x that denotes the ratio between the local rotational speed and the wind speed, axial induction factor in wake (a′) and the power optimization term $(x^2 dC_p/dx)$ where Cp denotes the power coefficient. Figure 4.12 depicts a typical wind turbine performance curve as a function of axial interference factor along with computed values of a and a′ parameters.

4.7 Summary

This chapter focuses exclusively on wind turbine blade design. Aerodynamics aspects and fluid dynamics considerations have been discussed in detail. Critical aspects of the blade theory have been described for the benefit of graduate students and blade design engineers. The performance of a wind turbine is strictly dependent on the number of blades attached to the rotor and their geometrical dimensions.

The blade theory states that the design of the outer contour of a blade must be based on aerodynamic considerations and the materials used in blade fabrication should provide the needed structural strength and stiffness under all operating conditions. This means that aerodynamic considerations and mechanical properties of the fabrication materials must be evaluated carefully during the design and development of a rotor blade. Proven models of actual flow conditions must be investigated for an engineer to predict aerodynamic performance of an airfoil accurately and under a variety of flow conditions.

The Bernoulli equation for a streamtube ahead of a propeller disc between a plane very far in front of the propeller and a plane immediately in front of it must be derived in terms of axial interference, tangential interference, drag and lift forces, undisturbed wind speed, drag-to-lift coefficient ratio, and relative wind speed. Expressions for the torque coefficient and power coefficient as a function of tip speed ratio, blade loading coefficient, flow angle, and drag-to-lift coefficient have been derived. This chapter also identified situations under which purely aerodynamic considerations may be modified by structural considerations to achieve safe and optimum blade performance under all speed environments.

Computed values of variations in pitch angle as a function of dimensionless radius, drag-to-lift coefficient, and blade tip speed ratio are provided. Beam theory applicable to blade design is discussed in detail, with emphasis on aerodynamic considerations and mechanical integrity under a variety of wind conditions. Fabrication materials and properties best suited for turbine blades are described, specifically in relation to mechanical strength and stiffness under extreme wind environments [6]. Mechanical integrity of the blades and turbine installations are critical under conditions of severe winds. Equations for computing deflection and bending moments at specific points along the blade length are provided. Important sources of loadings on rotor blades resulting from the earth's gravitational field and inertial and aerodynamic factors are identified and their effects on the mechanical integrity of blades and rotor stability are discussed in detail.

Mechanical strength and rotational stability of blades are vital and must be considered early in the design phase. Mechanical integrity of a rotor blade is dependent on its tensile strength, compressive strength, torsional strength, and stiff resistance of the materials used to fabricate the leading and trailing edges. Methods for maintaining mechanical stability and rotor operational safety have been recommended. In summary, rotor blade dimensions are strictly dependent on output capacity, wind speed, and operating environment in the vicinity of the installation site.

References

[1] A.J. Workman, *Introduction to Wind Turbine Engineering*, Butterworth, Boston, 1983, p. 57.
[2] M.O.L. Hansen, *Aerodynamics of Wind Turbines*, James, 1992, London, pp. 8 and 107.

[3] T. Wizelius, *Developing Wind Turbine Projects,* Earth Scan, Sterling, VA, 2007.

[4] J.F. Walker, *Wing Energy Technology,* 1996, John Walker & Sons, West Essex, U.K.

[5] A.R. Jha, Technical report: preliminary design aspects for wind turbine rotors, Jha Technical Consulting Service, Cerritos, CA, 1999, p. 7.

[6] C.A. Harper, Ed., *Handbook of Materials and Processes,* McGraw Hill, New York, 1984, p. 79.

Chapter 5

Sensors and Control Devices Required for Dynamic Stability and Improved Performance under Variable Wind Environments

5.1 Introduction

Suitable sensors and controlling devices are required to maintain wind turbine dynamic stability and enhance performance under variable wind conditions. Stall regulation, pitch control, and yaw control mechanisms will be described, with emphasis on reliability, safe operation, and improved turbine performance. Control mechanisms deploying micro-electro-mechanical systems (MEMS) and nanotechnology-based devices will receive serious consideration due to their fast response, accuracy, and minimum size and weight. For high-capacity wind turbines, the connection between the blades and the hub must be extremely rigid under variable wind environments if mechanical integrity is the principal design objective.

It is important to note that a teetering hub design reduces the loads on the turbine, thereby yielding improved dynamic stability. However, the advantages and

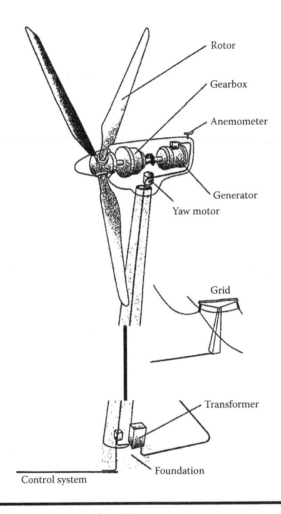

Figure 5.1 shows labels: Rotor, Gearbox, Anemometer, Generator, Yaw motor, Grid, Transformer, Foundation, Control system.

Figure 5.1 Horizontal axis wind turbine.

disadvantages of this hub design must be examined in terms of cost and design complexity. Monitoring sensors to record efficiency and turbine dynamic stability data must be collected randomly. Critical components such as rotor, generator, transmission system, and other components responsible for overall system efficiency, reliability, safety, and dynamic stability will be discussed in detail. Performance, reliability, structural integrity, and safety aspects will be investigated. Note that rotor dynamic stability is vital and depends on the number of blades and their airfoil characteristics.

Theoretical studies performed by the author indicate that two- and three-bladed wind turbines exhibit lower tip speeds and lower rotational speeds. High-capacity wind turbine systems such as horizontal axis wind turbines (HAWTs, Figure 5.1) require gear boxes to reduce the rotational speed to maintain dynamic stability under high tip speed conditions. Note that the tip speed (X) is the ratio between

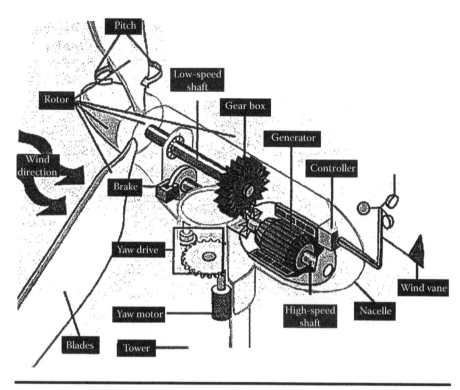

Figure 5.2 **Nacelle components.**

the rotor speed at the tip and the undisturbed wind speed. It is extremely important to mention that abrupt and ultrahigh rotational speeds are highly undesirable as far as the safety and dynamic stability of the turbine are concerned [1]. Most of the monitoring sensors and safety devices are located at the nacelle as illustrated in Figure 5.2. Critical elements of the wind turbine are located on the top of a tower as shown in Figure 5.3. Various techniques to reduce the rotational speeds under variable wind environments will be identified. Since high-capacity wind turbines will have longer blades and consequently higher towers, the system's dynamic stability and safety of both turbine and tower must be given serious consideration during the design phase. The author's aerodynamic studies of 2-D and 3-D wing designs indicate that the flow angle (φ), which is the sum of the angle of attack (α) and pitch angle (β), varies at different locations along the rotor blade length.

5.2 Regulation Control Systems

Regulation control systems [2] with fast feedback loops, microprocessor-based control mechanisms, and control algorithms are widely used by high-capacity wind

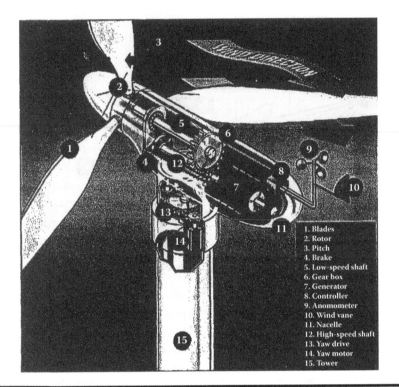

1. Blades
2. Rotor
3. Pitch
4. Brake
5. Low-speed shaft
6. Gear box
7. Generator
8. Controller
9. Anomometer
10. Wind vane
11. Nacelle
12. High-speed shaft
13. Yaw drive
14. Yaw motor
15. Tower

Figure 5.3   Critical elements at top of tower.

turbines to maintain dynamic stability and reliable operation under variety of wind conditions including turbulence. It is important to stress that both the pitch regulation and yaw control systems are required to achieve dynamic stability and safe and reliable operation of a rotor under a variety wind environments.

5.2.1 Pitch Regulation Control

Preliminary studies of dynamic stability indicate that pitch control is necessary for dynamic stability and reliable performance of a rotor. Pitch regulation control requirements indicate that the angle of attack must be kept below 5.7 degrees if optimum power coefficient and dynamic stability are to be maintained under a variety of wind environments. The flow angle (sum of the pitch angle and angle of attack) changes under variable wind speed conditions. The pitch angle must be adjusted at various locations along blade length to meet the flow angle requirements. The studies of a 30-m blade length reveal that both the flow angle and wind speed vary at locations along the blade length as shown in Table 5.1. The pitch angle must be adjusted by the pitch control mechanism comprised of a pitch motor, controller, and drive accessories to meet the flow angle requirements as summarized

Table 5.1 Pitch Angle Adjustments to Meet Flow Angle Requirements along 30-Meter Blade Radius

Radial Distance	*Wind Speed (m/sec)*	*Flow Angle (deg)*	*Pitch Adjustment (deg)*
Blade tip (R = 0)	60	6	0.7
0.8 R	48	7	1.7
0.6 R	36	9	3.7
0.4 R	26	14	8.7
0.2 R	12	27	21.7

in the table. The flow angle and pitch angle adjustments will vary as a function of blade radius or length. Note in general that the undisturbed wind speed will decrease to 67% in front of the rotor blade [2]. It is evident from the table data that the flow angle, wind speed, and pitch angle all increase from blade tip to hub location. As stated earlier, the values of these parameters change as a function of blade radius [2].

5.2.2 Description of Pitch Regulation Control System

The pitch regulation control system as illustrated in Figure 5.4 is the most critical wind turbine subsystem and is relatively expensive. It permits the blades to rotate about their radial axes as wind velocity changes during operation. It is therefore theoretical possible to have an almost optimum pitch angle adjustment at all wind speeds and at a relatively low cut-in wind velocity. Note that at high wind speed conditions, the pitch angle must be changed to reduce the angle of attack, which in turn reduces the aerodynamic forces on the rotor blades and maintains dynamic stability of the rotor. A reduction in aerodynamic forces on the rotor assembly will

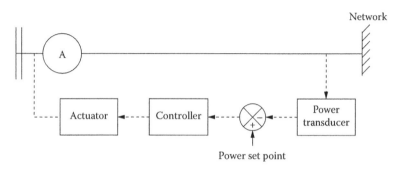

Figure 5.4 Pitch regulation system.

improve the dynamic stability of the entire turbine. Under these conditions, the mechanical power output from the rotor is limited to the rated power of the electrical generator. In some wind turbine designs, only the outer part of the blade is moveable and under this design configuration the aerodynamic load on the rotor is reduced compared to a stall-regulated rotor.

The pitch regulation system shown in Figure 5.4 is more expensive and is inherently more efficient and reliable than stall regulation. The figure is a block diagram of a widely used control system for pitch-regulated wind turbines. The critical elements of this regulation control system are the actuator, microprocessor-based controller, control algorithm, actuator, power set point device, and power transducer. The power generated by the wind turbine is measured using an electrical power transducer normally located in the base of the tower. The measured electrical signal is compared with the precision set point device and the error signal is then passed to the microprocessor-based controller. A control algorithm is used to obtain the required dynamic response for the actuator to rotate the blades in the desired direction. The rotation of the blades must be achieved smoothly and precisely.

Several types of actuators including hydraulic, electromechanical, electro-optical, and MEMS devices will be investigated with emphasis on the speed, dynamic range, and reliability of the response. The challenge or the difficulty is to change the pitch of the blades on a rotor that rotates constantly. The feedback loop of the control system must operate continuously to regulate the power above the rated wind speed. Under turbulent winds, it is common to encounter significant transient power fluctuations or excursions well above the set point, and this may jeopardize the dynamic stability of a turbine. To maintain dynamic stability in turbulent environments, a system to control yaw must be added in addition to a pitch control system.

5.2.3 Yaw Control System

The block diagram of a yaw regulation control system is similar to that of a pitch regulation control system, except yaw control is less sophisticated. As noted earlier, a pitch regulation control system is not enough to maintain dynamic stability under turbulence conditions. Implementation of yaw control is also essential to ensure dynamic stability and structural protection of a wind turbine system in a turbulent environment. Thus deployment of both pitch and yaw regulation systems adds to the cost and makes a system more complex, but the dual control system is necessary to main dynamic stability and safe operation of high capacity wind turbines regardless of wind speeds and turbulence environments.

A turbine rotor can be mounted on the upwind or downwind side of the tower. With downwind installation, the rotor blades can be coned outward to increase tower clearance. A downwind rotor configuration will align itself with the wind direction and thus eliminate the need for a yaw drive mechanism. However, downwind rotors suffer from increased noise and cyclic loads due to the tower "shadow" and are less common than upwind rotor designs. Furthermore, when operating under turbulent conditions, a downwind turbine rotor cannot provide the dynamic stability and reliable operation required.

5.3 Sensors for Monitoring Wind Parameters

Early wind turbine designs utilized wind vane and anemometer devices to monitor wind direction and speed. Accurate measures of these wind parameters are essential for optimum effectiveness of pitch and yaw control systems. MEMS and nanotechnology-based devices are available if fast response and large dynamic range of microprocessor-based control systems are the principal design requirements.

Early wind turbines deployed wind vane devices to monitor the wind direction. Modern direction monitoring devices for high-capacity wind turbines provide high accuracy and fast response. A direction monitor is located near the anemometer in Figure 5.2. Anemometers are widely used on wind turbines to record wind speeds. These devices are reasonably accurate, but nanotechnology-based devices are highly recommended, if fast and accurate wind response is a requirement.

Both these devices play key roles in providing instant wind parameters needed by the pitch and yaw regulation control systems. Without wind parameter data, these regulation control systems cannot operate. They are essential to maintain dynamic stability and reliable operation under varying wind environments including turbulence.

5.4 Transmission Systems

The mechanical power produced by a rotor is transmitted to an electrical generator by a mechanical system consisting of a gear box, perhaps a clutch, and a braking system to bring the rotor to a rest position in an emergency situation. The gear box is required to increase the speed typically from 20 to 50 revolutions per minute (rpm) to the 1000 to 1500 rpm required to drive most types of generators. The gear box is located at the top of the tower as illustrated in Figures 5.1 through 5.3. It may be a simple parallel-shaft box in which the output shaft is offset in a low-capacity wind turbine. However, output shaft design for high-capacity wind turbines is more sophisticated and expensive; it allows the input and output shafts to be in line, thus providing optimum compactness.

It is important to point out that shaft compactness and smooth shaft operation are the principal requirements for dynamic stability and reliable rotor operation. The two shafts in a gear box must provide vibration-free and reliable operation. Turbine designers must concentrate on high dynamic loads due to the fluctuating power output from a rotor under variable wind environments. Dynamic loads can be controlled by adding mechanical compliance and damping capabilities into the drive train mechanism. This technique is particularly important for high-capacity wind turbines that encounter high dynamic loads and the induction generator provides less damping. In summary, the performance of the transmission system depends on the efficiency, reliability, and compactness of the gear box components.

5.5 Electrical Generators

Several types of generators are commercially available for deployment in wind turbine systems. Induction generators, synchronous generators, variable-speed generators using fast transistor-based switching elements, and constant-frequency generators are commercially available. Generators fall into two categories: synchronous and asynchronous generators. Asynchronous generators are also known as induction generators and synchronous generators are generally known as alternators. Note all grid-connected wind turbines are coupled to three-phase AC generators to convert mechanical power to electrical power. The stators of both synchronous and asynchronous generators have three-phase windings on laminated iron cores to generate a magnetic field rotating at a constant speed.

The rotor of a synchronous generator has a field winding to carry a direct current. This field winding creates a constant magnetic field that interfaces into the rotating field created by the stator winding. Thus the rotor always rotates at a constant speed in synchronism with the stator field and the network frequency. In some generator designs, the rotor magnetic field is produced by permanent magnets that are not suited for large electrical generators.

5.5.1 Induction Generators

As stated previously, asynchronous generators are called induction generators. The design of the rotor of an induction generator is different from that of a synchronous generator. Induction generators and synchronous generators have non-rotating component called stators. Stators in both types of generators are connected to the network or load. Each stator consists of a three-phase winding on a laminated iron core to reduce noise and electrical losses in the core. The stator windings produce a magnetic field that rotates at a constant speed.

While both types of generators contain similar stators, their rotor configurations are very different. A synchronous generator rotor has a winding through which a direct current flows. The device is called a field winding because it generates a constant magnetic field that locks into the rotating field created by the stator winding. The rotor always rotates at a constant speed in synchronism with the stator magnetic field and the network or grid operating frequency. In some earlier low-capacity wind turbine designs, the rotor magnetic field was produced by permanent magnets. However, this was not a common practice for large wind turbines. The rotors in modern high-capacity turbines use the magnetic field windings to generate magnetic fields.

5.5.2 Rotor Configurations for Induction Generators

The rotor configuration of an induction generator is different from configurations of other generators. The rotor in this case consists of a squirrel cage of bars that are short-circuited at each end. There is no electrical connection onto the rotor assembly, and the rotor currents are induced by the relative motion of the rotor in conjunction with the rotating field of the stator winding. The rotor speed is exactly equal to the speed of the rotating field produced by the stator winding (as with a synchronous genera-tor). No relative motion is present and thus there is no induced current in the rotor. Therefore an induction generator always operates at a speed that is slightly higher than the speed of the rotating field produced by the stator winding. The difference between the two speeds is called the slip and equals about 1% during normal opera-tion. Although induction generators are not widely, millions of induction motors are currently in service throughout the world. Essentially an induction generator can be considered an induction motor with torque applied to the shaft rather than taken from it. Torque direction is the fundamental difference between an induction generator and an induction motor.

5.6 Performance Capabilities and Limitations of Synchronous Generators

Data covering design aspects and performance capabilities of electrical synchro-nous generators used in commercial power generating stations can be found in standard text books. These text books offer detailed information on the operat-ing principles, performance capabilities, and limitations of both synchronous and asynchronous generators. Brief studies performed by the author of the deployment of synchronous generators in electrical power plants reveal that most large electri-cal generators are synchronous and that synchronous generators are best suited for high-capacity wind turbine installations, because they are more efficient than induction generators. However, the major advantage is that their reactive electri-cal power flow is easily controlled. In a synchronous generator, the direct current flowing in the rotor field winding magnetizes the rotor. By increasing the rotor field current, the reactive component of the electrical power generated can be exported or transmitted to a power grid or load network. Similarly, by reducing the rotor field current, reactive power going to the network can be reduced. Thus, controlling the flow of reactive power gives a firm control over the voltage of a power system. In this way, constant voltage can be maintained at variable speeds (Figure 5.5) and irrespective of the electrical load environments.

It is important to mention that most high-capacity wind turbines belong to the HAWT category. The rotor of a HAWT produces torque pulsations at the frequency with which the rotor blades pass the tower structure. If these torque pulsations are at the natural frequency of the spring-mass system created by the blades and their

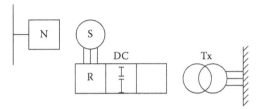

Figure 5.5 Schematic diagram of a variable-speed wind turbine generator.

network connection via the synchronous generator, resonance frequencies and some very large oscillations of the drive train will occur. This occurred with one early wind turbine. It is possible to provide damping of the drive train with mechanical devices such as fluid couplings, but this technique increases design complexity and system cost. Because of problems caused by great oscillation, synchronous generators are not normally used in fixed-speed wind turbine installations [3].

5.7 Critical Rotor Performance Parameters

Several types of rotors were introduced commercially to fit certain applications and power ratings. Figure 5.6 illustrates popular early rotor design configurations. Dutch type rotors with four blades were popular for certain applications. They demonstrated power coefficients as high as 0.17, high torque capability, low revolution per minute

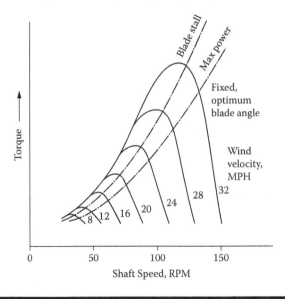

Figure 5.6 Early rotor designs.

rates, and inefficient blade designs. They were followed by farm rotors with power coefficients close to 0.30, high torque capability, but low rpm rates. The high energy losses meant poor efficiencies. In the early 1950s, modern propeller rotor design was introduced. It demonstrated a maximum power coefficient of 0.47, low torque capability, high rpm levels, and efficient blade design involving exotic airfoil designs.

5.7.1 Rotor Design Classifications

Most rotors were classified according their potential applications for HAWTs or VAWTs. Upwind rotors and downwind rotors were recommended for HAWT installations, while the Savonius, Darrieus, and Giromill rotors were suggested for VAWT installations because of their unique performance characteristics. Rotor performance is measured in terms of power coefficient (C_p) as a function of blade tip speed-to-wind speed ratio (X). Typical power coefficients of various rotors are summarized in Table 5.2. The table data indicate rotor performance over a number of tip speed- to-wind speed ratios and clearly show that propeller rotor configurations offer optimum performance in a variety of wind environments. Most modern high-capacity wind turbines deploy three-bladed rotors to achieve optimum performance over a wide range of wind conditions.

5.7.2 Dynamic Stability and Structural Integrity

Attempts will be made to recommend various techniques to determine and record the dynamic stability and structural integrity of a rotor. These two characteristics are the most important ones for a rotor. The safe and reliable operation of a wind turbine depends on them.

Wind turbine designers suggest that a two-bladed rotor configuration is least expensive, but it runs faster, possibly causing serious problems from tip noise, blade erosion, teetering to limit fluctuation forces as the rotor blades sweep varying wind

Table 5.2   Power Coefficients of Various Rotors

Rotor Type	Optimum C_p	Range of Tip Speed-to-Wind Speed Ratio (X)
Savonius	0.3	0.80 to 0.85
Dutch four-arm	0.14	2.0 to 3.0
Darrieus	0.32	5.5 to 6.5
Two-blade	0.43	4.5 to 6.5
Propeller (ideal)	0.55	3.0 to 7.0

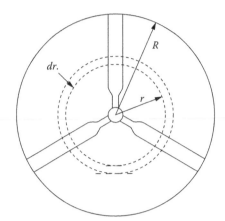

(a) Wind turbine rotor configuration with three blades

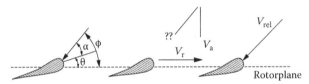

(b) Cascade of aerofoils in the radial direction from the axis

Figure 5.7   Airfoil characteristics.

velocities, and dynamic instability. In addition, with a teetered hub, the rotor is hinged at the hub in such a way as to allow the plane of rotation of the rotor assembly to tilt backward and forward a few degrees away from the mean plane. Under such conditions, the rotor will experience serious dynamic instability, leading to vibrations of high magnitude.

The rocking motion or teetering of a rotor at each revolution significantly reduces the loads due to gusts and wind shear. Most high-performance wind turbines deploy three blades to achieve reliable and safe performance with no compromises in structural integrity and dynamic stability. Furthermore, a three-blade rotor configuration as shown in Figure 5.7 provides better efficiency and improved reliability than a two-blade rotor. A three-blade rotor will cost about 6% more than a two-blade rotor. In summary, a three-blade rotor offers improved rigidity, stiffness, fatigue strength, and structural integrity under most wind environments except turbulence.

Rotor performance depends completely on airfoil characteristics of the blades (Figure 5.7). Essential characteristics include insensitivity to roughness, low noise production, excellent stall characteristics and high lift-to-drag ratio for improved efficiency under varied wind environments. Rotor performance is somewhat dependent on airfoil shape and its aerodynamic characteristics. Improved aerodynamic performance requires thinner blades toward the tip, tapers in chord and twists. The taper, twist, and improved airfoil characteristics of the rotor blades should combine to achieve the best possible wind energy capture for the rotor speed power and site conditions.

Note that the blades of an upwind side rotor will be coned onward to increase power clearance. Conversely, a downwind rotor will align itself with the wind direction, thereby eliminating the need for a yaw drive system under moderate wind environments. However, downwind rotors suffer from increased noise and cyclic loads due to the tower "shadow." Because of design complexity and unfavorable performance under gusty wind environments, upwind rotors are preferred by the wind turbine designers over downwind rotors.

5.7.3 Monitoring of Stress Parameters

Random monitoring of bending moments, shear forces, and blade tensile and compressive stresses using MEMS and nanotechnology-based devices will keep a turbine operator informed of the structural health of the blades. Particularly, the tensile and compressive stresses generated under gusty and turbulence environments must be carefully monitored because they can affect the dynamic stability, turbine reliability, and structural integrity of the rotor. Deployment of these monitoring devices on remotely located wind turbines can play a significant role in monitoring structural health and system reliability in gusty and turbulence environments.

5.7.4 Stall Regulated Rotors

Stall regulated wind turbines deploy a rotor configuration in which pitch angle distribution along the blades is constant at all wind speeds. Under stall conditions, the angle of incidence onto the rotor blades increases. The lift forces are reduced due to stall, but the drag forces rise and the increase in rotor power that would otherwise occur is prevented. This results in large aerodynamic forces on the rotor. Stall regulation is simple and does not require a sophisticated control system. However, rotor designers experience some difficulty in designing stall regulated rotors. Active research and development activities are continuing on both three-dimensional and dynamic stall effects.

5.7.5 Factors Affecting Wind Energy Capture and Turbine Performance

Several factors can affect wind energy capture and ultimately deteriorate turbine rotor performance:

- Orientation relative to wind vector if a turbine is not omnidirectional [3]
- Wind speeds at which the turbine starts and stops

- Design wind speed if the turbine operates in a power-limited mode
- Performance sensitivity to off-design conditions
- Rotor response to transient factors such as wind gusts or turbulence

From a practical view, rotor performance can be estimated from the ratio of total energy extracted by the rotor to the total wind energy available. In other words, the measure of rotor performance is strictly dependent on the maximum power coefficient that can be achieved under the defined operating environments. It is important to mention that most high-capacity wind turbines are designed to supplement the power from a utility grid system and are thus forced to operate at fixed rotational speeds to generate alternating current at the designated grid frequency. The use of converters and inverters is discouraged by the high cost of the equipment and consequent reduction in efficiency. Because of the variability of the wind, constant rotational speed rotors are forced to operate at off-design conditions frequently. If the angular speed was an entirely free design parameter, a propeller-based wind turbine could be operated at its peak power coefficient regardless of wind environments and the centrifugal stress variations in the blade angle of attack.

It is extremely important to mention that transient conditions affect the dynamic stability of a rotor under gusty or turbulence conditions and may pose a serious threat to rotor reliability and the structural integrity of the tower. In addition, rotor blades experience higher magnitudes of bending moments and shear forces as illustrated in Figure 5.8 if effective pitch and yaw are not controlled immediately. Furthermore, under transient conditions, catastrophic failure of an entire wind turbine system cannot be ruled out. At times, a complete shutdown of a wind turbine under uncontrollable transient conditions should be considered to prevent personal injury and property damage.

In the case of a wind turbine with small or moderate capacity, a teetering hub rotor could be advantageous because it reduces the fluctuating loads on the turbines that affect the dynamic stability of the rotor. In case of high-capacity wind turbines, large Coriolis forces, centrifugal forces, tensile stresses, compressive stresses, and transient loads on the rotor must be avoided to maintain stability, reliability, and structural integrity of the rotor. As mentioned earlier, the wind speed will be optimum at the tip of the blade and minimum at the root of the blade. Thus, the bending moments and centrifugal forces at the tip are significant. This indicates that "beefing up" of the tip section must be given serious consideration to meet dynamic stability and structural integrity requirements [4].

Sensors capable of monitoring the centrifugal forces, Coriolis forces, tensile stresses, compressive stresses, and shear stresses at strategic locations along the axial blade length must be calibrated to obtain accurate data on the structural health of the rotor and blades. Excessive bending moments and shear forces as illustrated in Figure 5.8 can pose serious threats to the structural integrity of the blades and dynamic stability of the rotor, particularly, under gusty wind and turbulence environments. Monitoring devices or sensors are recommended to determine the magnitudes

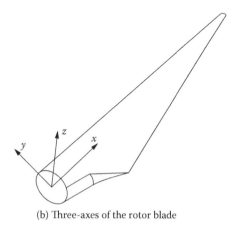

(a) Bending moment (M) and shear force acting on rotor blade (T)

(b) Three-axes of the rotor blade

Figure 5.8   Bending moments and shear forces.

of the bending moments and shear stresses for high-capacity wind turbines. Such sensors are neither required nor justified for small or moderate capacity wind turbines to minimize system cost and complexity. As stated earlier, a wind turbine generates unpleasant environmental noise levels and there is no need to monitor them.

5.8 Impacts of Airfoil Characteristics on Blade and Turbine Performance

It is interest to mention that airfoil characteristics (Figure 5.7) play an important role in the performance and reliability of rotor blades. Essentially a wind turbine blade is a beam of finite length with airfoils as cross-sections to create a pressure difference between the lower and upper sides of the wind and give rise to lift. At the tips of the blades are leakages that allow air flows around the tips from the lower side to the upper side. In other words, the streamlines flowing over the wing will be deflected inward and the streamlines flowing under the wing will be deflected outward. These conditions create a jump in the tangential velocity at the trailing edge followed by a continuous sheet of stream-wise vorticity in the wake behind a wing. These sheets are known as trailing vortices.

Classical aerodynamics principles demonstrate that a vortex filament of certain strength can model the flow past an airfoil for small angles of attack so long as the angle of attack does not exceed 5.7 degrees regardless of wind environment. This is because the flow for small angles of attack is mainly inviscid and governed by the linear Laplace equation. Under this assumption, an airfoil can be replaced by one vortex filament of finite strength and the lift produced by 3-D wind can be modeled for small angles of attack using a series of vortex filaments (bound vortices) oriented in the span-wise direction of the wing.

In summary, blade performance is based on the vortex filament orientation and trailing vortices. However, the performance will not improve if rotating blade Coriolis and centrifugal forces are not controlled at angles of attack exceeding 5.7 degrees. At large angles of attack (more than 12 degrees), the air stream cannot stick to the airfoil surface, thereby creating a turbulence effect under which airfoil begins to stall. Aerodynamic scientists state that an airfoil is most efficient when the angle of attack remains below 7.5 degrees. However, most designers prefer to keep the angle of attack below 7.5 degrees for other reasons. Airfoil design and the relative thickness of the blade must be given serious consideration to achieve stall-free and reliable rotor performance under various wind conditions. The circumferential force must be kept small compared to the thrust to retain the high relative speed needed to produce optimum blade power.

5.8.1 Vortex System behind High-Capacity Turbine

A HAWT functions as a high-capacity wind turbine. The rotor of a HAWT consists of a number of wing-shaped blades. If a cut is made at a radial distance r from the rotational axis, a cascade of airfoils can be observed as shown in Figure 5.7. The figure illustrates the rotor of a three-blade HAWT with blade radius R. The local angle of attack is given as α by the pitch of the airfoil θ, the axial velocity is V_a, and the rotational velocity at the rotor plane is denoted by V_{rot}. The cascade of airfoils can be seen at the bottom of the figure. The vortex system produced is located behind the HAWT. If the lift and drag coefficients are known, it is easy to compute the force distribution for the airfoils along the blades. The blade element momentum (BEM) method can be used to compute the axial induction factor, the longitudinal induction factor, and the various loads acting on the turbine. The Biot-Savart equation can be used to calculate the induced velocities in the rotor [4].

5.9 Automatic Shut-Down Capability

Automatic shut-down capability of a wind turbine in an emergency situation created by gusty winds or turbulence should receive serious consideration for safety reasons. Since no operational or maintenance personnel are usually near a wind turbine installation, a built-in provision for automatic shut-down is preferred to prevent damage to the turbine and nearby structures. Brief studies of turbine safety and reliability performed by the author indicate that a HAWT with a downwind design configuration could provide a built-in automatic adjustment to wind direction, serving as a safe operational feature under most wind environments. However, this adjustment is not possible when abrupt changes in wind direction occur. Under abrupt wind direction changes, a rotor with three or more blades could offer smooth and stable performance. Under turbulence conditions, there is no design provision for automatic shut-down.

5.10 Critical Design Aspects of HAWT and VAWT Rotors

Comprehensive aerodynamic studies of wind turbines performed by the author reveal that most high-capacity systems operate as HAWTs because of their superior performance and improved reliability. A HAWT offers higher output power, relatively lower environmental noise levels, and higher power coefficients. VAWTs tend to stall under gusty wind conditions and present serious dynamic stability problems. A VAWT requires lower installation height, leading to inferior operation at lower wind conditions, reducing both output power and efficiency. On the other hand, a VAWT has small blade length that prevents large bending moments and deflections and offers significant savings for the tower structure and accessories. The environmental noise level from a VAWT is lower than that from a HAWT because of low installation height.

5.10.1 Techniques for Improving Turbine Reliability and Performance

Techniques to capture maximum energy under low wind speed conditions should be explored. As explained earlier, wind energy capture depends on wind speed. Low wind speed means low energy capture; a wind turbine operation will not be cost-effective under such condition. Studies by aerodynamic scientists reveal that a variable generator excitation control technique implemented in turbine design will realize significant improvement in the amount of wind energy capture and consequently higher wind turbine power output in low wind speed environments. These scientists further note that his technique also improves the aerodynamic rotor control to ensure safety and reliability of the entire system. Turbine efficiency depends on the number of blades used and the blade efficiency. Blade performance is a function of flow angle, wind speed, blade radius, angle of attack, blade pitch angle, and axial and longitudinal interference factors. In summary, turbine performance is based on rotor and blade efficiency.

Ethernet interface can provide continuous monitoring of turbine performance under low wind conditions. It is interesting that residential wind turbine systems deploying twisted high-efficiency blade designs offer improved turbine performance, optimum angle of attack along the entire blade length, and unique aerodynamic design features. Implementation of charge controllers would offer better system reliability and improved turbine performance.

Scientists at MIT claim that a new concept based on jet engine principles offers hope for improved wind turbine design with significant improvements in efficiency and reliability. The new design concept forces the air around instead of through the rotor blades. In other words, the system works by channeling the wind into a vortex that spins the blades and generates mechanical power that can be converted into electrical power via a generator.

Various augmentation schemes are available to enhance the performance of wind turbines. Most effective designs demonstrating significant improvement in performance include concentration, aerodynamic lift, and diffusers and ejectors [4]. However, performance improvement based on these schemes involves a trade-off of additional weight, cost, and increased tower design complexity. Trade-off studies are required to determine performance improvement versus the increases in weight, cost, and complexity before an augmentation system is selected.

Optimum values of critical design parameters that may achieve significant performance improvements must be given serious consideration. For example, optimum wind speeds must be within the recommended range throughout the year at an installation site. Furthermore, the maximum power coefficient—the main performance indicator of a wind turbine—requires wind speeds of 20 to 30 m/min, a drag-to-lift coefficient ratio less than or equal to 0.01, and an angle of tilt close to 5.7 degrees. The impacts of drag-to-lift coefficient ratio and tangential speed ratio (X) on power coefficient can be seen from the data in Table 5.3 indicating that the power coefficient continues to increase as the tangential speed ratio increases at the lowest value of drag-to-lift coefficient ratio. However, the power coefficient increases initially, then decreases. Note that the power coefficient values decrease rapidly at the highest drag-to-lift ratio as the tangential speed ratio increases. It is important to mention that higher power coefficients are possible at a drag-to-lift coefficient of 0.01, regardless of the tangential speed ratio.

In summary, both the drag-to-lift coefficient and the tangential speed ratio significantly influence the power coefficient or efficiency of a wind turbine. It is interesting to mention that greater power is available at higher wind speeds excluding gusty and turbulence environments.

5.10.2 Sensors for Ensuring Efficiency, Dynamic Stability, and Structural Integrity

An installation site must be selected carefully, with consideration given to the adverse effects of temperature inversion and rapid variations in wind speeds. Turbine installations at sites frequently subjected to gusty winds and turbulence are undesirable

Table 5.3 Impacts of Drag-to-Lift Ratio and Tangential Speed Ratio on Turbine Power Coefficient

Drag-to-Lift Ratio	Tangential Speed Ratio				
	1	*2*	*3*	*4*	*5*
0.00	0.52	0.56	0.57	0.58	0.586
0.01	0.50	0.55	0.56	0.55	0.542
0.03	0.46	0.47	0.45	0.42	0.371
0.05	0.45	046	0.45	0.41	0.372
0.07	0.43	0.44	0.39	0.34	0.314
0.09	0.41	0.38	0.32	0.25	0.185

if dynamic stability and structural integrity are required. Critical installation site selection parameters should be carefully examined and evaluated prior to installation of a high-capacity wind turbine.

Temperature inversions cause a stratification of the atmosphere near the surface that is detrimental to efficient operation of a wind turbine. Installations where variations in wind speed and direction exceed 25% from one year to the next should be avoided because optimum capture of wind energy is not possible. It is important to mention that wind speed and direction are subject to change at any time for any location. Rayleigh distribution and Weibull distribution techniques should be used to estimate the wind speed for a potential site. Meteorological scientists state that Rayleigh distribution provides reliable estimates to project annual power generation from small and medium capacity systems. Reliable sensors must be deployed to monitor wind speed distribution because it plays a critical role in determining the power generated.

Most small and medium capacity wind turbines use anemometer devices to monitor wind speed with moderate accuracy. MEMS and nanotechnology-based monitoring sensors now available offer reliable and accurate information on wind speed. A monitoring device must be installed at an appropriate location on a nacelle where it will be free from interference from nearby objects.

Wind speed and direction data monitored by the sensors is sent to the controller for transmission to other sensors or subsystems. The controller essentially directs the yaw motor to turn the rotor to face toward or away from the wind, depending on wind direction. Wind data is also sent to the braking mechanism that directs the rotor to slow down when necessary. Under certain operating environments such as turbulence, the braking mechanism will send a command to stop completely to avoid structural damage to the tower and injury to maintenance personnel nearby.

5.10.3 Blade Twist Angle Adjustments

Adjustment of blade twist angle is necessary to optimize performance in variable wind environments. It is impossible to stop a rotor and then adjust the twist angle. As a matter of fact, adjustment of twist angle is very tricky while a turbine is in operational status. An operator must be highly experienced and make incremental adjustments accurately and quickly. A sudden or incorrect adjustment could affect dynamic stability of a rotor, producing violent vibrations that could damage the tower structure. In summary, only an experienced operator can adjust blade twist angle to maintain optimum rotor performance.

5.11 Low Harmonic Content Electrical Generators for Improving Efficiency

To achieve optimum generator efficiency, it is necessary to convert all the DC power into AC power through an inverter with an operating frequency compatible with grid frequency requirements. This enables a generator to operate over a wide range of wind speeds without disturbing the grid frequency. In addition, all the modern high-capacity wind turbines deploy high-speed transistors as switching devices to generate waveforms much closer to sine waves with fewer harmonic distortions as shown in Figure 5.5. The generator design concept shown offers high efficiency leading to improved system output.

Further improvement in generator efficiency may be realized by cooling the stator windings via a forced-air system operating at a specific rate and cooling temperature. Cooling of stator windings improves generator performance and increases reliability. Forced-air cooling also keeps the graphite exciter brushes and the rotor windings at safe operating temperature. Cooler temperature increases the safety and reliability of the rotor and stator windings over extended durations. When the operating temperature is kept constant at an optimum value, the electrical losses and IR losses in the windings are significantly reduced, leading to more generator efficiency and improved reliability.

5.11.1 Bearing Reliability

Lubrication is absolutely essential for bearings used by gears, turbines, and generators operating at high rpm levels. Lubrication promotes smooth and reliable operation, particularly at high rpm conditions. A gear box in a wind turbine system is located between the rotor and an electrical generator that rotates typically between 4000 and 5000 rpm. The bearings supporting the rotor assembly and generator must be kept well lubricated with an oil that meets the room temperature viscosity requirements recommended by the American Institute of Mechanical Engineers or an equivalent authority on temperature specifications to ensure high reliability and maintenance-free operation. The gears are completely immersed in the oil to maintain efficiency, minimize wear and tear, and ensure reliable operation. Oil level in a gear box must be checked regularly to maintain quality and optimum quantity. No scheduled maintenance requirements are recommended by wind turbine suppliers or installers. As long as the operating wind conditions are smooth and free from turbulence effects, wind turbines can run efficiently over very long periods without maintenance.

5.12 Impacts of Loadings on Structural Integrity of Wind Turbine

The structural integrity of a wind turbine can be compromised by loadings. Determining stresses generated by various loading sources plays important role in fatigue analysis of a system. As mentioned previously, the most important sources of loadings are gravitational forces, inertia, and aerodynamics.

5.12.1 Impact of Gravitational Loading

The gravitational field of the earth causes a sinusoidal loading on rotor blades and the stress may eventually lead to structural damage. When a rotor blade is down-rotating, the trailing edge is exposed to tensile stress and the leading edge is exposed to compressive stress. Similarly, in the up-rotating position, the trailing edge encounters compressive stress while the leading edge meets tensile stress. At the same time, edge-wise bending moments may develop. This means that rotor blades under gravitational loading conditions may be subjected to tensile stress, compressive stress, and bending moments as illustrated in Figure 5.9. The stresses and moments can seriously undermine the dynamic balance of a rotor and structural integrity of a turbine. Bending moments of higher magnitudes will disturb the dynamic balance of rotor and lead to performance degradation. It is important to mention that wind turbines are designed to operate 20 years or longer. Assuming a 25-rpm turbine speed and an operating life of 20 years, a wind turbine will be subjected to 2.6×10^6 (20 years × 365 days × 24 hours × 60 minutes × 25 rpm) stress cycles from gravity. Since modern wind turbine blades are more than 30 m long and may weigh several tons, the stresses from gravitational loading are critical for fatigue analysis of the rotors and assessment of the structural integrity of a wind turbine.

5.12.2 Impact of Inertial Loading

Preliminary aerodynamic studies by the author indicate that inertial loading occurs when a turbine is accelerated or decelerated under abrupt change in wind speed. Inertial loading can also occur when a braking torque is applied to a rotor under

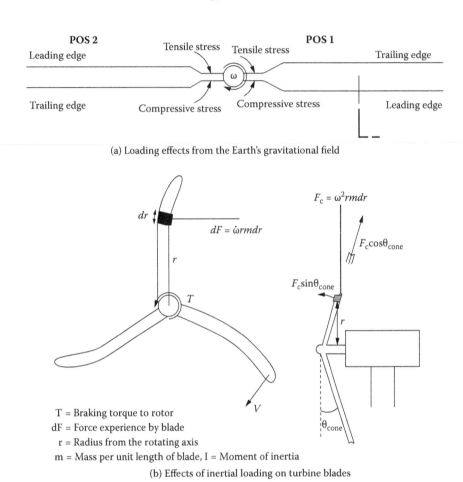

(a) Loading effects from the Earth's gravitational field

$F_c = \omega^2 r m dr$

$dF = \dot{\omega} r m dr$

$F_c \cos\theta_{cone}$

$F_c \sin\theta_{cone}$

T = Braking torque to rotor
dF = Force experience by blade
r = Radius from the rotating axis
m = Mass per unit length of blade, I = Moment of inertia

(b) Effects of inertial loading on turbine blades

Figure 5.9   Inertial loading.

certain operating conditions. Under certain braking conditions, a small section of the blade will feel a force of specific magnitude (dF = dw/dt r m dr) in the direction of rotation, where dw/dt is the rotation speed, r is the radius of the rotating axis of the blade section, m is the mass per unit length of the blade, and dr is the blade element length. The braking torque (T) applied to the rotor shaft is equal to the product of the moment of inertia of the rotor and the rate of rotating speed of the rotor. Note that inertia loads can also stem from local acceleration that may contribute additional force acting on a rotor. Inertial loading of greater magnitude can stem from excessive centrifugal force acting on the blades, which is generally applied to reduce the flap-wise bending moment (Figure 5.9).

The centrifugal force acting on an incremental section of a rotor blade at a radius r from the rotational axis can expressed as F_c which is equal to w² r m dr, where m is the mass of the incremental section of the blade and w is the angular velocity of the rotor. Due to coning, the centrifugal force has a component in the span-wise direction of the blade, but the normal component of centrifugal force yields a flap-wise bending moment in the opposite direction to the bending moment produced by the thrust, thereby reducing the total flap-wise bending moment. It is interesting to visualize how blade coning influences bending moments in the blades, which can impact the stresses on the blades and their dynamic balance under higher stress conditions.

5.12.3 Impact of Aerodynamic Loading

The theories of aerodynamics and fluid dynamics indicate that aerodynamic loading is caused by air flow past a turbine structure including the rotor blades and tower. The wind field seen by a rotor varies in space and time due to the atmospheric factors including turbulence. It is important to mention that the wind field seen by a rotor is strictly characterized by shear [4]. In other words, the mean wind speed increases with height above the ground. To maintain a neutral stability condition, this shear can be estimated with an error not exceeding ±10%. Wind shear is defined as a terrain surface characteristic and is expressed in meters of roughness length (Z_0). The Z_0 value varies from 10^{-4} m over a water surface to 1 m over city surfaces. Roughness lengths and other terrain surface characteristics are summarized in Table 5.4 [4].

It is desirable to note that wind shear is given as a sinusoidal variation of wind speed seen by a rotor blade with a frequency corresponding to the rotation speed

Table 5.4   Roughness Lengths of Terrain Surface Characteristics

Terrain Surface	Roughness Length (m)
City	1.0
Forest	0.8
Surface with trees and bushes	0.2
Farmland with closed appearance	0.1
Farmland with open appearance	0.05
Farmland with few buildings	0.03
Bare soil	0.005
Snow white	0.001
Smooth sand	0.0003
Water	0.0001

of the rotor. Furthermore, the turbulent fluctuations superimposed on the mean wind speed also produce a time variation in the wind speed and thus in angle of attack value. To simulate the behavior of a wind turbine using an aero-elastic code, it is necessary first to generate a realistic wind field and remember that a tower also contributes to variations in inflow. Variations in the inflow differ for upwind and downwind rotor configurations.

The turbulent inflow and sinusoidal variations from wind shear make yaw regulation system design more difficult. However, a wind turbine may operate under a yaw condition if the direction of the wind is not measured correctly or the yaw control system malfunctions. In this case, the reduced wind speed at the rotor plane has a component normal to the rotor axis and a tangential component to the rotor. If the blade at the top position corresponds to a zero degree angle, the rotor moves in the same direction as the wind and the relative rotational speed is reduced at the bottom position corresponding to 180 degrees. In summary, the efficiency of a yaw regulation control system is based on sinusoidal variations of the wind speed due to shear and the roughness length of the terrain surface characteristics.

5.13 Summary

This chapter focuses on the sensors and devices that ensure the efficiency, dynamic stability, and structural integrity of various elements of a wind turbine. Sensors to provide the pitch and yaw control capabilities are described, with emphasis on reliability and safety. Design aspects of pitch regulation and yaw control systems are summarized and factors such as time response and reliability are detailed. Pitch angle adjustment is defined to meet flow angle requirements at various locations along the length of a 30-m blade. Methods for improving dynamic stability and structural integrity of a rotor are discussed.

Monitoring sensors incorporating MEMS and nanotechnologies that provide accurate data on wind speed and direction are described. Anemometers currently installed on wind turbines lack accuracy and fast response. Various techniques have been described to improve the efficiency and reliability of mechanical transmission systems and methods for improving the efficiency and reliability of a generator are outlined. Forced-air cooling for stator windings is described; it improves the efficiency and reliability of an electrical generator. Techniques to maintain high speed bearing reliability are also explained.

The structural integrity of a rotor and a complete turbine system may be compromised by loadings generated by various sources. These loadings play an important role in fatigue analysis of a system. The most important sources of loadings include gravity, inertia, and aerodynamics. The adverse effects of these loadings on the dynamic stability and structural integrity of a rotor are identified. Aerodynamic scientists and wind turbine designers note that the most serious impact can be expected from gravitational loading, because earth's gravitational field creates sinusoidal loading on the blades. This loading can generate tensile and compressive stresses of high magnitudes, leading to dynamic instability and structural damage. It can also generate bending moments that can further complicate the dynamic balance of a rotor. Rotor efficiencies for various design configurations are mentioned as functions of tip speed ratio.

The impacts of airfoil characteristics on rotor performance and turbine efficiency are discussed in detail. Air leakage occurring at the blade tips may cause a sudden jump in the tangential velocity at the trailing edge, thereby producing trailing vortices that can disturb the angle of attack. Aerodynamic scientists state that the most efficient airfoil is a critical parameter if high efficiency and reliability of a rotor under various wind environments are required.

Sinusoidal variations in wind speed observed under turbulent conditions may generate a shear effect that adversely impacts both dynamic stability and angle of attack. Wind shear is defined as a terrain surface characteristic and is generally expressed in meters as roughness length (Z_0) that varies from 10^{-4} m over water to 1 m over a city surface. These values indicate that shear effect on turbine performance and dynamic stability of the rotor is most serious in a city environment. Preliminary studies by the author reveal that optimum wind turbine performance requires a wind speed of 20 to 30 m/min and a drag-to-lift ratio equal to or less than about 0.01.

References

[1] A.J. Wortman, *Introduction to Wind Turbine Engineering*, Butterworth, Boston, 1983, p. 57.
[2] J.F. Walker and N. Jenkins, *Wind Energy Technology*, 2002, John Wiley & Sons, Chichester, p. 45.
[3] Ibid., p. 47.
[4] M.O.L. Hansen, *Aerodynamics of Wind Turbines*, 2nd ed., Earth Scan, London, 2007, p. 8.

Chapter 6

Stand-Alone Wind Turbine Systems

6.1 Introduction

Stand-alone wind turbine systems play an important role where utility-based power systems are nonexistent. Stand-alone wind turbine generators are best suited for powering remote sites. These light power plants deploy small or moderate capacity wind turbines and banks of batteries sized to carry electricity for household consumption during severe winter winds and summer emergency periods. Studies performed by the author reveal that the high cost, poor reliability, and frequent maintenance requirements of early stand-alone power systems discouraged many customers, but modern wind turbine technology has solved most of the problems and people in remote locations beyond the reach of commercial utility power lines accept the stand-alone power system concept.

The studies further reveal that reliable and efficient inverters, state and federal tax incentives, and widespread availability of low-cost photovoltaic (PV) or solar panels have revolutionized stand-alone wind power systems. Data compiled by Pacific Gas & Electric Company indicate that the number of stand-alone wind-based power systems continues to mushroom at a rate of 29% annually [1]. The utility's data further indicate that the market for such systems will continue to expand in areas not served currently by commercial utility companies. Aggressive research and development focused on critical components have demonstrated significant improvement in cost, efficiency, and reliability of stand-alone power systems.

Implementation of solar cell technology into stand-alone wind turbine systems offers a cost-effective and practical choice for customers living in remote areas that have no commercial utility power lines. It is important to mention that remote stand-alone power systems that depend solely on wind turbines may require back-up electrical generators or high-capacity battery systems. The modularity of photovoltaic technology transformed the remote stand-alone power system market by enabling homeowners to tailor power systems to their needs and budgets.

The use of wind turbines in stand-alone applications required trade-off studies in terms of cost, power consumption needs, and availability of electrical energy at all times. Various configurations of stand-alone wind turbines can be seen in Figure 6.1. Historically wind turbines of moderate capacity (known as wind chargers or windmills) were used to charge batteries of remote power systems typically deployed to pump water and grind grain. Today, wind machines are frequently used to drive AC motors directly, without the need for inverters for specialized pumping applications. In addition, wind machines can be used to generate heat for residential and other applications.

The author's studies of stand-alone power systems reveal that the electrical energy must be generated near the site of use to minimize distribution cost and transmission losses. On farms, small wind turbines are ideal for this purpose. Paul la Court in Denmark was one of the first people to connect a windmill to an electrical generator and gave a tutorial course for "agricultural electricians." He installed in his school one of the first wind tunnels in the world to investigate rotor aerodynamic principles. Gradually, however, diesel engines and steam turbines took over

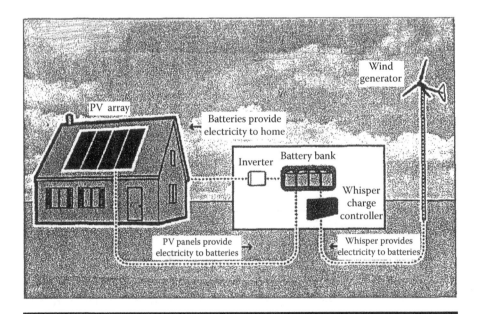

Figure 6.1   Stand-alone wind turbine system.

the production of the electrical energy. However, during the two world wars when fossil fuel was scarce, wind turbine technology was considered as a source for generation of large amounts of electrical energy.

6.2 Historical Background: Use at Remote Sites

Besides mechanically pumping water, wind turbines are best known for generating electrical power at remote sites for decades. During the1930s, when only 10% of wind farms generated electricity for homes, thousands of small wind turbines were used for irrigation and pumping water. Small turbines are widely used in rural areas to supply electricity to remote areas where power lines are not available.

Wind turbine planners claim that as many as 110,000 small stand-alone wind turbines are used by nomadic herdsmen in remote regions of Northwestern China. These stand-alone wind turbines are small enough to be carried on horseback from one encampment to another. They represent the sole sources of electrical power available on the great Asian Gobi Desert that stretches from China to Russia. According to suppliers of stand-alone wind turbines, China alone builds approximately, 10,000 wind machines per year.

What is a remote site? The rules are not strict. Commercial utility companies will build a power line almost anywhere if someone pays for it. For one or two customers at a remote site, a utility cannot justify the cost of running lines. Stand-alone wind turbine installers feel that as a rule anyone residing more than 1 km from an existing power line will find it cheaper to install an independent power system rather than connect to a commercial utility power line. For this reason, about 75% of all small wind turbines built are destined for stand-alone systems at remote sites. Some find their ways to remote locations in Canada and Alaska. Others serve mountain-top telecommunications sites where utility power lines do not exist. Surprisingly, increasing numbers of stand-alone power systems are installed to serve the lower 48 United States. Some homeowners determined to generate their own electrical energy even though they could just as easily buy it from local utilities.

It is important to point out that the wind is an intermittent source of energy; remote applications generally require some form of alternate energy source. For remote residential that require a fairly steady supply of electricity, battery storage becomes a necessity. Batteries store surplus electrical energy generated on windy days for later use. Battery systems require only direct current (DC) for charging, but DC can be converted to alternating current (AC) similar to power from a commercial utility source.

Studies performed by the author indicate that a remote installation site requires an average wind speed greater than 9 mph to generate the electricity from a stand-alone power system or small wind turbine at lower cost than power from gasoline-based or diesel generators. At such sites, according to energy planners, a small wind turbine may be more cost-effective than a photovoltaic system alone.

However, battery-charging wind systems as illustrated in Figure 6.1 will be found more practical and cost-effective.

6.3 Configurations of Stand-Alone Systems

Potential design configurations of hybrid stand-alone power system with back-up capability and variable speed propellers are shown in Figure 6.2. The propeller generally deploys two rotor blades (As shown) to achieve optimum economy and design simplicity.

6.3.1 Hybrid System with Back-Up Capability

This design for a hybrid stand-alone power system consists of a low-capacity wind turbine, solar array as a renewable energy source, batteries to be charged, an inverter to convert DC to AC electricity, battery charger, back-up generator, and AC load center. This configuration of a hybrid stand-alone system can simultaneously charge the batteries and provide AC to power devices operating at 110 V, 60 Hz frequency. The hybrid system [1] offers a cost-effective approach and ensures continuous supplies of electricity if any of the two power modules becomes temporarily inactive. In other words, if the turbine cannot operate because of abnormal wind conditions, solar panels will provide the DC electricity for charging the batteries and AC

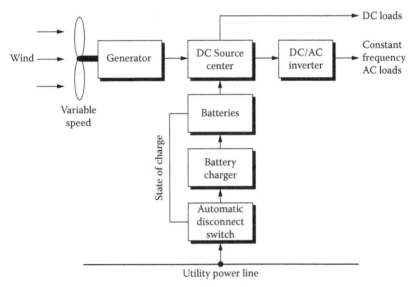

Figure 6.2 Hybrid stand-alone system with back-up capability and variable speed propellers.

electricity power to drive appliances. Conversely, if solar power is not available due to heavy clouds or rains, electrical energy is available from the small wind turbine.

If you wanted a wind turbine for a remote site, a dealer will meet your electrical energy requirements and provide you with batteries of appropriate specifications to handle the required power load. A solar dealer would provide solar panels that would cover your roof with PV cells or modules with optimum power ratings. Market survey date indicate that Bergey Wind Power and Northern Power Systems—small wind turbine pioneers—have successfully demonstrated wind and solar hybrids that utilize the assets of both technologies. In many remote sites, wind and solar resources complement each other: the strong winds in the winter are balanced by strong sun in the summer, thereby enabling the hybrid system designers to reduce the size of each component [1]. System designers have found that these hybrids perform even better in terms of reliability and efficiency when coupled with small low-cost back-up generators to reduce the number of batteries needed; fewer batteries will provide significant savings in procurement costs.

6.3.2 Micro Wind Turbines

Micro wind turbines with very low capacities are best suited for charging the batteries on sailboats and recreational vehicles. Such small wind turbines that output fewer than 100 W are commercially available and effective for charging batteries with minimum cost and complexity. These micro wins turbines are so small that they can be erected or taken down by a single person in a short time. Micro wind turbines can used to charge automobile batteries and power security lighting systems in remote locations. PV technology can be used to charge batteries, but the cost of a solar power system and possible low sunlight levels favor the micro wind turbine for certain conditions; they can be installed for less than $2000.

6.3.3 Applications of Micro Wind Turbines

Some applications of low-power wind turbines do not require storage capabilities. One example is cathodic protection of pipelines. A micro wind turbine provides an electric charge to the surface of a metallic pipe to counteract galvanic corrosion in highly reactive soils. Furthermore, storage is not needed during calm winds because corrosion is a slow process occurring over long periods. When the wind returns, it protects the exposed metallic surfaces from galvanic corrosion. This protection is best suited for remotely located pipes carrying oil, gas, water, and other nontoxic liquids. At one time all cathodic protection in rural areas was provided by wind turbines that generated few watts. Today most pipelines are protected by small PV modules, but in cloudy environments this protection may not be effective.

Another application of micro wind turbine technology is charging automobile batteries to power security lighting in rugged and isolated areas. It is interesting to mention that PV modules are currently deployed extensively for such applications because of lower installation costs. There are few technical reasons why micro wind turbines cannot be used if the installation cost exceeds $1500.

Some micro wind turbines have been designed to perform several tasks simultaneously, e.g., charging electric vehicle batteries, providing emergency lighting, powering security sensors, and generating electric power compatible with utility company specifications. Some micro wind turbine are available capable of supplying power for all-electric modern homes located in remote areas where utility power lines do not extend.

6.3.4 Micro Wind Turbines for Rural Electrification

It is important to mention that about one-third of the world's people live without electricity. Many people in villages of India, China, Arabic countries, and African countries lack access to utility power. Many third world nations are scrambling to expand their power systems to meet demands for rural electrification. While China which has embarked on a program of rapid industrialization, about half of the population lived without access to utility power as recently as 15 years ago. Most countries are following the pattern set by the United States during the 1930s: building new power plants and extending transmission and distribution lines from the cities to rural areas. However, this approach to rural electrification no longer makes sense since reliable hybrid power systems comprised of wind turbines and solar power systems became commercially available.

Developing nations' installation of hybrid power systems is more cost-effective than stretching the service areas of heavily loaded and unreliable generating plants in large cities. The new hybrid power systems generate very little power in comparison to central power plants, but the third world uses very little power because most of its population cannot afford to buy and operate electrical appliances due to high power costs. To demonstrate this point, 1 kWh (one unit) of electricity provides ten more essential services in India than it provides in Indiana.

Hybrid power systems that feature small or micro wind turbines because of their sufficient power output and low procurement cost. Local governments using micro wind turbines can provide electricity to villages quickly in emergency situations. As the electrical loads on central power systems expand to remote villages, a compact hybrid stand-alone system as shown in Figure 6.3 can be moved easily to provide power to remote villages; thus, rapid electrification of remote villages is possible at minimum cost and with simplicity. This compact wind turbine power system is fully capable of providing all the electricity for a home while feeding excessive electricity to commercial utility grid power lines. This compact stand-alone system has higher generating capability than a micro wind turbine and can be designed in any configuration shown in Figure 6.1 to suit a specific application.

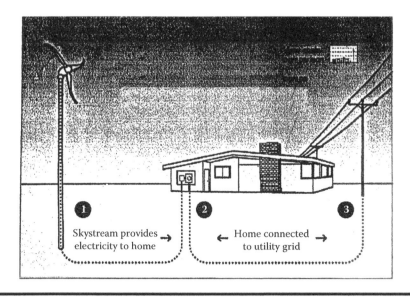

Figure 6.3   Compact hybrid stand-alone system.

6.3.5 Power Generating Capacity

Despite their advantages, wind turbines are less modular than PV power modules. When scaling up outputs from remote power systems, it is easier to add more solar modules to a PV array than it is to add wind machines. Although the output of micro wind turbines is less than that from most PV modules (typically 50 W), each additional wind turbine requires a separate tower and control system, thereby adding installation and procurement costs. Wind turbines are sized from 50 to 250 W. Recent models have power generating capabilities approaching 500 W. In brief, PV modules can be added in smaller increments than micro wind turbines [2].

6.3.6 Telecommunications Applications in Remote Areas

It is difficult and economically infeasible to install power lines for telecommunications applications in remote and hostile areas where hybrid versions of micro wind turbine power systems can play a critical role. Advances in solar and wind technology have eliminated the need to install expensive stand-alone power systems because remote systems do not usually depend totally on a single source. For example, on an overcast winter day when PV generation is low due to poor solar intensity, sufficient wind energy is probably available to compensate for the loss of solar power.

Hybrid power systems consisting of wind and solar power modules also permit use of smaller, less costly components than would be needed if a system design configuration depended on only one power source. This design flexibility can significantly lower the installation and procurement costs for a remotely located power source. Furthermore, the designer of a hybrid power source is not required to size the components for worst-case operating conditions by specifying a larger turbine and storage battery bank than necessary.

A hybrid design could include a small and inexpensive fossil-fueled back-up generator for the same reasons. In effect, a stand-alone system can replace the fuel and the maintenance expenses of a back-up generator with a larger solar and wind turbine combination. However, this approach would require trade-off studies to analyze performance, reliability, and cost. Depending on generator size and overall power consumption, an electrical generator can "top up" discharged batteries and meet electrical needs not met by a combined wind and solar generation scheme.

6.3.7 Cost Reduction Techniques

The author's studies of hybrid versions of power sources author indicate that despite significant improvements in reliability, performance, and size, initial procurement costs of power systems remain high. One way to control system cost is to reduce the demand for power. In addition, latest technologies must be explored to improve the energy efficiency, allowing users to meet their energy needs from smaller and less costly power systems. The studies further indicate that little in the way of lifestyle sacrifice is required because modern compact fluorescent lights and energy-efficient appliances need less power and less expensive wind turbines.

6.3.8 Reducing Power Demands

Reducing electrical energy consumption through conservation and efficiency measures will certainly reduce the demand for electricity, thereby allowing the deployment of inexpensive, low-capacity power systems. Such reductions will permit users to stretch each kilowatt hour to do as much as work possible. This is particularly true at remote locations where individuals must spend time, effort, and money to generate electricity. In addition, knowing how, where, and when electrical energy is used is more important for a stand-alone power system than for an interconnected wind turbine. One cost saving idea is to determine what appliances will be used at a remote site and estimate their power consumption in an effort to reduce operating times for sensors and appliances that do not contribute to system reliability and longevity.

6.3.9 Typical Energy Consumption by Electrical Appliances

To size any electrical power system, one must estimate energy consumption by appliances and other devices requiring electricity. After the total need is established, it is easy to design a wind, solar, or other energy providing system. The typical energy amounts consumed by common appliances according to Pacific Gas

& Electric of California are summarized in Table 6.1. The electrical appliances are categorized in terms of heating, air conditioning, and other household functions.

It is important to note that a stand-alone wind turbine in a remote location will not have to support many of the appliances cited in the table. A remote location will most likely require a 50-W lighting source, a security sensor not consuming more than 100 W, and a small fan with a 50-watt rating. Therefore, the total energy consumption at a remote location may not exceed 200 W. However, remote energy consumption will certainly increase if other appliances are added to the load. The average energy consumption of a residence is about 5 kW if a window air conditioning unit, a microwave oven, two fans, six 60-W bulbs, a refrigerator, computer and printer, radio, and a dishwasher operate simultaneously. Note that most high-wattage appliances such as toasters, washing machines, dryers, microwave ovens, central air-conditioning units, and electric ranges usually do not operate simultaneously. Therefore, sizing of a stand-alone power system should be based on overall power consumption by appliances and the number of hours they operate.

6.3.10 Techniques for Reducing Energy Consumption

It is always more economical to save energy than to generate it with a hybrid power system comprising of micro wind turbine and a solar power source. As mentioned earlier, hybrid power systems are best suited for remote and unmanned locations. It is extremely important to determine the minimum energy consumption needed for the services expected if optimization of the value of a renewable power system and minimization of procurement costs are the principal deciding factors. Among alternate methods for reducing electricity consumption, for example, is powering a hot water heater or stove with gas or other fuel. However, installation of a gas line to a remote location is not economically feasible. Similarly, heating with gas, oil, propane, or wood is far more economical at a remote site than heating with electricity. It is wise to consider all available options to save energy consumption at remote sites. Electric cooking stoves consume great amounts of energy and exert high peak power demands that determine the sizes of inverters and other hybrid components. Cooking with gas or propane instead of a microwave will save power and provide economical service.

6.4 Stand-Alone Systems for Remote Sites

This section discusses energy requirements for remote sites and emphasizes economic feasibility and reliability. Studies of stand-alone power sources performed by California's Pacific Gas & Electric Company found that remote power generation must be confined to lighting, refrigeration, and water pumping uses because remote sites are seldom served by municipal water sources. Compact fluorescent lights can

Table 6.1 Typical Power Consumption by Household Appliances

Appliance	Consumption (W)
Heating	
Space heater	1500
Heat pump	4000
Air Conditioning	
110-V window unit	1500
220-V window unit	2500
Central unit	4500
Water Heating	
Electric heater	350
Heat pump	200
Refrigeration	
10-cu ft manual defrost	50
20-cu ft frost-free	145
Laundry	
Electric dryer	5000 per load
Washing machine	250 per load
Lighting	
Bulbs	60 to 100
Other Electrical Appliances	
Microwave oven	750 /1500
Ceiling fan	200/300
Portable fan	100/200
Ceiling fan	200/250
Air cooler	250–300
Toaster	750/850

**Table 6.1   Typical Power Consumption by
Household Appliances (Continued)**

Appliance	Consumption (W)
Iron	1000/1400
Laser printer	1100
Facsimile	120/150
Personal computer	50/150
Black-and-white television	75
Color television	225
Plasma television (52 in.)	400

significantly reduce the electricity demand. Consumption may be further reduced by lighting only areas where light is needed and by simply turning off lights when not needed. These techniques can cut lighting costs by two thirds. Power reduction techniques are particularly critical for remote and unmanned operating sites.

6.4.1 Refrigeration Appliances

Significant energy savings can be achieved with refrigeration appliances. Modern refrigerators consume close to 1000 W; today's energy-efficient models consume 700 to 800 W watts in typical American homes. Sun-frost refrigerators exhibit the highest efficiencies (300 W) but cost about $2500. A standard refrigerator may push an energy budget over the edge and require an additional wind turbine or more solar panels to meet the demand. Sun-frost devices may be well worth the added cost— another trade-off to consider when designing a stand-alone power system.

6.4.2 Air Conditioning Units

Depending on climate and house size, an air conditioning unit can double the energy consumption of an otherwise energy-efficient home. If air conditioning is necessary, a trade-off study should be conducted to determine whether an evaporative cooler or air cooler can provide a comfortable temperature [2]. Swamp coolers use far less energy and work well in arid climatic environments such as the Southwestern United States. Energy ratings of electrical appliances including washing machines and refrigerators are available from the *Consumer Guide to Home Energy Savings* published by *Home Energy* magazine which also offers tips on saving energy without reducing comfort. It is important to mention that some behavior modification may prove beneficial for optimizing the performance of a hybrid power system. For example, minimizing energy-intensive discretionary electrical loads on days when power is reduced extends battery life and stores extra power for use when needed for important functions such as water pumping or additional refrigeration. It is desirable to learn to synchronize weather and lifestyle, e.g., wash laundry on windy and bright sunny days to save electrical energy without affecting lifestyle.

Turning off unneeded appliances is a very small burden for energy-conscious consumers. Those who cannot compromise a comfortable lifestyle may face a rude awakening. It is advisable to reduce your electric consumption until you are ready to make a transition to producing your own electrical power. An average American household should able to reduce its consumption to about 3600 kWh per year or about 10 kWh per day. Most Europeans live comfortably with that much electricity or less. A reduction of power consumption will determine the size of a power system that will meet your needs, whether wiring should be for AC or DC , and the required voltage for operating a hybrid power system.

6.4.3 Selection of AC or DC Operation

After the power consumption requirement is defined and AC or DC power is selected, performance specifications for the hybrid system components can be established, after which the procurement of the components and installation of turbine and tower can be completed. The author's studies of stand-alone hybrid power systems reveal that they can generate and store DC electricity. PV arrays produce DC energy; wind turbines generate AC energy that is rectified to DC energy so it can be stored in batteries. A back-up generator provides constant frequency AC energy needed by electrical appliances. It is important to mention that the addition of a back-up generator as shown in Figure 6.2 provides high reliability and greater design flexibility in sizing a hybrid stand-alone system, but the back-up generator will increase system complexity and cost.

The two major aspects of a stand-alone system are power generation and storage. Since most electrical loads will be supplied by the batteries most of the time, the choice becomes whether to feed the load with DC directly or with AC though an inverter as illustrated in Figure 6.1. The figure clearly identifies the inverter and storage batteries. The efficiencies of electronic inverters exceed 90% but inverters add cost and complexity to a system. Conventional homes use AC electricity to operate appliances. If owners of stand-alone systems with inverters want to add commercial utility power or sell their properties, they do not need to rewire from DC to AC. Furthermore, new houses are always wired for AC. Such homes are easier to finance than those wired for DC. Banks are sometimes reluctant to finance homes with stand-alone power systems and prefer to finance homes with conventional wiring. DC wiring is justified for refrigeration and water pumps because these appliances limit inverter losses. It is interesting to mention that sun-frost refrigerators are built in both DC and AC versions.

6.4.4 Parameters of Generating System

Appropriate values of electrical parameters of a generating system must be defined to maintain operational continuity and reliability. Output voltage levels for solar panels, wind turbines, batteries, and a back-up generator system must be specified. Note that low-voltage operations require thicker and heavier cables or wires to carry same amount of electrical power as high-voltage systems. For the same sizes of cables, low-voltage operations waste more electrical energy in resistance losses than higher-voltage operations. Thus the voltage preferred for operating a remote hybrid power system safely and efficiently is directly proportional to the amount of power that passes through it. This is an important design consideration for a hybrid stand-alone power system. In brief, the selection of operating voltage is strictly dependent on power consumption. According to the editor of *Home Power*, a 12-V operation is preferred for a building requiring fewer than 2 kWh per day. Homes consuming power below 6 kWh per day should use 24-V service; those requiring more than 6 kWh per day should have 48-V systems.

The author's studies of wind turbines indicate that some micro wind turbines operate at 12 V because the power output is only few watts. Most wind turbines and solar hybrids use 24 or 48 V, depending on the power output of the solar arrays and how the arrays are connected. A voltage of 120 is considered appropriate when the power consumption approaches 20 kWh per day. The initial cost for a 120-V system is extremely high. Furthermore, high-voltage systems do not permit greater design flexibility for renewable power systems than they do for 24- or 48-V systems. If you need to move a solar array into a more exposed position or move a wind turbine to a hilltop [3], a 120-V power system may be the only choice for minimizing transmission line losses and cable costs. In brief, operating voltage for a hybrid power system must be selected yield maximum economy and minimum cable losses.

6.5 Sizing of System Components

For a small load powering one transistor radio and couple of 20-W bulbs, a 12-V PV system consisting of one or two 50-W solar panels with DC output makes the most cost-effective choice. For applications such as vacation cabins, a PV power system is easier to work with and less expensive than a micro wind turbine. However, if the load increases above a few hundred watts, wind and solar hybrid power schemes become more attractive and practical. For loads exceeding 4 kWh per day, a wind turbine is far more cost-effective than PV modules. Small PV arrays typically cost about $25 per kilowatt hour. Large amounts of electrical energy can be generated from small wind turbines at lower cost. For example, a wind turbine with a 3-M rotor can generate electricity in excess of 3 kW and will cost about $2000. A wind turbine with a 7-m rotor can generate more than 20 kW and will cost about $10,000; errors in such cost estimates may exceed ±10%.

6.5.1 Sizing and Performance Capabilities of Solar Arrays

Sizing of solar arrays is dependent on several factors including the latitude of the installation location, peak sun hours per day, height of the array, solar intensity, climate conditions, and type of sun tracking scheme deployed. To estimate the potential electric generation from a PV array that may contain hundreds of 2-W solar modules, multiply the rated capacity of the PV array by the peak sun hours per day for the installation site region. For example, a 200-W solar array will generate about 1000 W per day in a region with five peak sun hours per day—equivalent to 365 kWh per year. Deployment of a tracking mechanism for the solar array can achieve about 45% more energy than a fixed array and may boost the annual solar energy close to 530 kWh. Specific details on sizing solar arrays are available in *Solar Cell Technology and Applications.**

As noted earlier, output of a solar power system depends on several factors, while output of a wind turbine depends on relatively few: wind speed and direction, rotor dimensions, and tower height. However, after a wind turbine is installed, its output depends totally on wind parameters. After the roof panels are installed for a solar system, the power output depends on the peak solar intensity (hours per day), temperature of the array surface, and the shadowing effects from the panels. Finally, the power output estimate from a solar installation will be less accurate because solar installations are more subject to parameter changes than wind turbines.

6.5.2 Sizing and Performance Capabilities of Inverters

Inverters convert DC energy into AC energy. Several types have been developed, but the electronic versions are best suited for solar-based power systems. A few years ago, the efficiency of an electronic inverter ranged from 88 to 91%; it has now significantly improved close to 95%. Compact commercial electronic inverters yield efficiencies ranging from 92 to 94%. It is important to point out that cost, reliability, and efficiency are the most critical design parameters of the electronic inverters for applications in solar power modules.

Inverters operate conventional appliances, whether power is generated by a PV array, wind turbine, or hybrid power system. Most inverters produce a modified AC sine wave that can serve a variety of AC loads ranging from standard electronic components (computers and stereos) to washing machines. In cases of sensitive electronic components, inverters with low harmonic distortions are preferred to maintain optimum performance and reliability of the devices.

---

* A.R. Jha. Taylor & Francis, 2010, Boca Raton, FL.

To determine the size of an inverter, it is necessary to calculate the electrical load demand in watts from all the appliances that are likely to operate simultaneously. The inverter should be sized to handle the surge requirements of induction motors in refrigerators and washing machines that operate over extended periods. Small appliances often demand 1.5 to 2 times their rated operating currents when they start. This fact must be considered during the design and development phase to ensure the stated reliability and longevity specifications are met. Large appliances such as washing machines and refrigerators can draw three to four times their rated operating currents when they are switched on. For example, a 500-W electric motor induction requires 1500 to 2000 W for start-up. Such power surge requirements are easily met with wind turbines, but present a problem for solar-based power modules. All inverter manufacturers list their continuous and surge outputs, but the continuous output rating represents what the inverter can supply over a long period without intermittent failure, overheating, or performance degradation. Few loads operate continuously, and those that do draw very small current. Sandia National Laboratories recommends sizing an inverter to 125% of the expected continuous load to avoid performance degradation and component overheating.

For safe operation, a 2-kW inverter should run most minor appliances or electrical loads, while major appliances such as washing machines and microwave ovens must be operated singly. A 5-kW inverter will be required to operate a washing machine, refrigerator, well pump, and microwave oven simultaneously. A microwave oven, hair dryer, or iron will draw an individual load from 1000 to 1500 W, but they operate only for short times. They may influence the size of the inverter needed, but they contribute very little to total energy consumption. Stand-alone power system designers predict that an inverter with a capacity of 5 to 8 kW is required if all the appliances discussed above operate simultaneously.

An inverter should also provide fuse protection from the various sources of generation and power factor correction for inductive loads. In addition, the inverter design should have provisions for disconnect switches on both the AC and DC sides and power management switches to limit certain loads from exceeding its capacity.

It is not advisable to operate appliances such as electric dryers, electric hot-water heaters, or electric stoves on a stand-alone power system unless you are using them as dumps for excess energy generation. These electric appliances consume large amounts of electricity and place unreasonably high current demands on an inverter. Propane gas is a better answer for powering such appliances provided they operate for short durations or only in emergencies.

Some inverters are designed to provide battery charging functions. A separate inverter should be available for this purpose and it should be capable of handling multiple inputs via hook-ups to renewable sources and from a back-up generator. Without a battery charger, there is no way to ensure that the batteries stay fully charged to provide electrical energy if needed.

6.5.3 Sizing and Performance Parameters of Batteries

Two distinct types of batteries are common in hybrid stand-alone power systems: lead–acid and nickel–cadmium batteries. In addition, batteries used in trucks and golf carts can be used if their charge–discharge cycles, capacities over extended periods, and reliability under uncontrolled environments are acceptable for use in hybrid stand-alone power systems. Sustaining a rated battery charge under various temperature environments and over extended periods must be given serious consideration. Sealed batteries are preferred to achieve reliability and longevity over the long term. Lead–acid batteries have higher current capacities, but must be maintained to ensure full energy capacity at all times. Their capacity is significantly reduced at lower temperatures and thus present a maintenance problem. As to cost, lead–acid batteries are less expensive than nickel–cadmium batteries. A nickel–cadmium battery with the same capacity (ampere hours) as a lead–acid battery could cost as much as $1500; the cost of a lead–acid battery is significantly lower. Nickel–cadmium batteries are lighter, more reliable, and do not require maintenance.

It is a good policy to separate batteries from inverters, switches, and service panels if longevity and reliability of the batteries are the principal requirements. In addition, this practice prolongs the lives of the electrical components while guarding against the ignition of hydrogen emitted by sparks during high-voltage charges. Nickel–cadmium batteries are more costly than lead–acid batteries, but offer other advantages. They are best suited for cold climates because they are less susceptible to freeze damage, require very little maintenance, and last longer than lead–acid batteries. Nickel–cadmium batteries are widely used in computers and other electronic devices and can be fully discharged without harm to electronic circuits or devices. Lead–acid batteries are best suited for power applications and do not suffer from "memory effects."

Correct sizing of batteries is critical to ensure constant performance and reliability of the operating components of a stand-alone power system. It is advisable not to undersize the batteries, inverters, and wiring to avoid catastrophic failure of the critical elements. A lead–acid battery bank has no capability for expansion if storage is insufficient after few years of operation. More batteries cannot simply be added in small increments. Essentially, a lead–acid battery bank is a fixed-size entity. Expansion requires a complete new bank of batteries of the proper voltage wired in parallel. Suggested locations for solar panels, batteries, inverters, controllers, and fused switches are shown in Figure 6.4. A vent should be provided to remove acid fumes from the storage area. If possible, lead–acid batteries should be housed in a well-ventilated room separated from living quarters to ensure safety of the occupants.

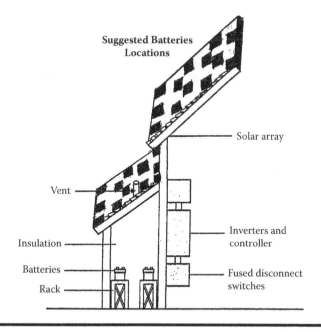

**Suggested Batteries
Locations**

Solar array

Vent

Inverters and
controller

Insulation

Batteries

Rack

Fused disconnect
switches

Figure 6.4   Components of power storage system.

6.5.4 Sizing and Performance Parameters of Solar Panels

Sizing of solar panels is relatively simple. Panels can be added to compensate for losses to a back-up generator. In other words, any power shortage in the back-up generator capacity can be made up by installing additional PV modules to an array or install a larger micro wind turbine.

6.6 Stand-Alone Systems with Utility Power Back-Up

Potential design configurations for stand-alone wind turbine power systems are illustrated in Figure 6.5. The figure shows variable speed propellers, locations of the inverters and batteries, water pumping apparatus, electrical generator with DC or variable frequency AC capability, and application for thermal loads. Hybrid versions of stand-alone power systems incorporating micro wind turbines and photovoltaic sources along with reliability and performance issues were described in Section 6.5. A stand-alone power system with utility back-up capability is useful when a utility is reluctant to implement a direct connection or when homeowners want to wean themselves gradually from a commercial utility grid system. Electronic controls are necessary to monitor continuously the performance of a battery bank. If the batteries become depleted or completely drained, an automatic switch closes the

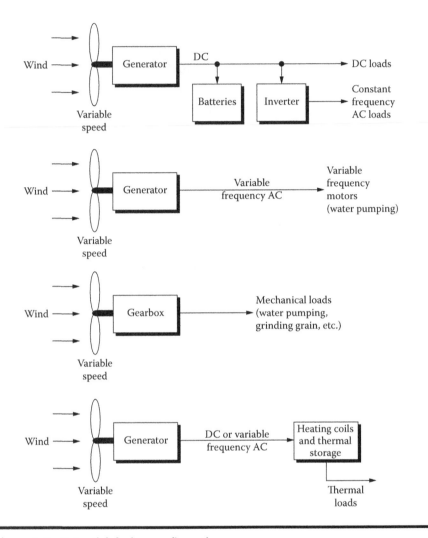

Figure 6.5 Potential design configurations.

utility power line to the battery charger, thus starting the charging process until the batteries are fully charged.

If renewable power sources consistently fail to provide the necessary power and a commercial utility is nearby, it may be desirable to engage utility back-up power to charge the batteries, in essence substituting utility grid power for a back-up generator. The independent power source (not the utility company) still handles all the electrical loads and continues to provide service to all electrical loads if the utility power is interrupted. This utility back-up power approach makes a lot of sense in regions where utility power service is not reliable.

It is important to mention that utility back-up is like the uninterruptable power sources widely used to protect computers from power outages regardless of cause. Many critical computer systems are required to operate continuously from batteries. The utility source is used only to charge the batteries. The electrical loads draw current from an inverter that is constantly fed by the batteries.

6.6.1 Economic Aspects of Remote Systems

In general, users of remote power system do not expect their wind turbines and solar power generation to compete with the cost of commercial utility-generated electrical energy. Such users typically install remote power systems because the cost of extending utility power lines to their sites is significantly higher. As a rule of thumb, California utility companies charge new customers about $10 per foot for overhead power line extensions [2] and charge 50% more for buried power cables. Pacific Gas & Electric charges about $60,000 per mile to extend power lines. Sometimes utility companies offer discounts for educational institutions and religious buildings. The situation in Europe is slightly different. The French electric companies charge rural customers as much as $32,000 per mile to bring utility lines to rural customers.

Under these financial conditions, a stand-alone system could pay for itself in one year if its owner resides more than 1 km from utility power lines. For someone considering a stand-alone power system in the United States strictly on economic grounds, it would be cheaper to bring in a utility power line if the site is fewer than 1000 ft from existing lines. The material presented above clearly establishes the economic principles for extending commercial power lines in various situations.

6.6.2 Cost Analysis for Stand-Alone Hybrid System

A cost analysis for a hybrid version of stand-alone power system comprising of a small micro turbine and a compact solar system is based on the following facts. The micro wind turbine assumes a rotor diameter of 3 m and turbine capacity equal to 13,000 W. The output of the solar array is assumed to be 800 W. A hybrid system allows downsizing of battery storage requirements. Both the solar and wind systems are priced based on minimum power storage: 13 kWh for 2 days. Using sophisticated design techniques, we can eliminate one inverter, thereby realizing a significant reduction in procurement cost as illustrated by Table 6.2. The back-up generator selected is sufficient to handle the electrical loads for a power system of this size. Output power rating, power available per day over a 5-hour duration, storage capacity, and procurement costs of both systems are summarized in the table. The values of the performance parameters quoted are based on extensive engineering analysis and judgment.

Table 6.2 Procurement Costs and Performance Parameters of Stand-Alone Systems

Parameter	Solar Power Module	Micro Wind Turbine
Power output rating of system (W)	800	1500
Power output per day over 5 hr (kWh)	4.0	8.2
Storage capacity (kWh)	13.0	13.0
Procurement cost ($)	11,600	10,000
Back-up generator/charger capacity (W)	6000	

Summary:

Refrigerator procurement = $500

Duplicate inverter = (– $1400)

Total power output = 8300 W

Total power consumption requirement = 12.2 kWh

Total storage requirement = 26 kWh

Total system procurement cost = $25,000

6.6.3 Cost Estimate for Extending Line from Existing Utility

This section presents a cost estimate analysis for extending a utility power line located 2500 ft from a customer residence. As mentioned earlier, commercial utilities charge $10 per foot to extend a power line to a customer's premises alone with additional fees summarized in Table 6.3. The assumed parameters and conditions are:

> Cost of service for 2500 foot line = $25,000 with rate escalation of 10%
> Minimum energy purchase requirement = $1200
> Annual kilowatt hour purchase order = 4500 with inflation rate of 5%
> Utility company charge for electricity = 10 cents per kilowatt hour

6.7 Stand-Alone Wind Turbine–Based Systems for Various Applications

This section discusses potential applications of stand-alone wind turbine power systems with emphasis on performance, reliability, and operational capability in severe wind environments. Performance capabilities and limitations of stand-alone

Table 6.3   Estimated Cost of Extending Utility Power Line

Service Year	Service Cost ($)	Utility Power Purchased ($)	Net Payment ($)
1	25,000	1200	26,000
2	00	1200	1200
3	00	00	1200
4	00	00	1200
19	00	2503	2503
20	00	2753	2753

Total amount paid to utility company over 20 years of service = $54,035.

Net present value of payments due utility company after 20 years = $38,500

wind turbine systems for telecommunications, village electrification, and building heating and water pumping applications will be summarized.

6.7.1 Telecommunications

Telecommunications operations demand reliability and uninterrupted energy sources. Wind turbines deployed by the telecommunications industry encounter more extreme weather, particularly in isolated rural environments, and operate continuously over extended periods (sometimes in excess of 7500 hours per year). These unattended power sources are expected to function for much longer periods of time than home-based power systems and even commercial wind power plants. The author's preliminary studies of wind turbines for telecommunications applications reveal that reliability and minimal maintenance are the principal design requirements. According to designers, only robust wind turbines using fully integrated, direct-drive designs perform satisfactorily in the rugged and isolated environments characterized by telecommunications sites as shown in Table 6.4. Wind turbines made by various companies listed in the table have operated at several sites and the users are impressed by the reliability of the machines even when operating at an average wind speed of 6 m/sec or 11.66 knots.

6.7.2 Performance Capabilities of HR 3 Hybrid

The HR 3 hybrid system designed by Northern Power Systems in the United States is widely used for telecommunications applications. This power system in operation installed in Antarctica to handle telecommunications worked continuously for 12 hr during a fierce Antarctic storm. This demonstrates the mechanical integrity and uninterrupted performance capability of the HR 3 wind turbine. However, the

Table 6.4 Continuous Load Capabilities of Reliable Wind Turbines

Wind Turbine Designation	Continuous Operating Load (kW)
NPS HR 1 (hybrid from Northern Power Systems)	0.4
Bergey 1500 (wind turbine from Bergey Wind Power)	0.4
NPS HR3 (hybrid from Northern Power Systems)	1.0
Bergey Excel (wind turbine from Bergey Wind power)	3.5

radio station temporarily went off the air when the exhaust pipe stack for the back-up generator blew away during the storm. When the storm started to abate, the HR 3 resumed its normal operation, recharged its storage batteries, and fully restored all the station's telecommunications. Twice during the first 2 years of operation, high winds at the site blew the anemometers away after they recorded wind speeds of 126 mph or 56 m/sec. Since then, this installation site has endured even stronger winds. The performance of the HR 3 was so impressive that three more HR 3 wind turbines were installed at the site.

Two U.S. manufacturers, Northern Power Systems and Bergey Wind Power, are the most reliable suppliers of wind turbines suited for telecommunications applications. Their wind turbines meet the telecommunication industry's principal requirements, namely reliability and low maintenance. The turbines are capable of meeting the continuous electrical load requirements of telecommunications installations encountering average wind speeds close to 6 m/sec or 14.5 mph. Other components of these systems include a PV array and an 84-kWh battery bank that significantly decreases the operating time of a diesel generator, yielding a significant saving for costly diesel fuel for a generator. Alternate fuels such as propane or gasoline can also operate generators.

6.7.3 Fuel Savings from Use of Hybrid Systems

Based on 15 years of operation as shown in Table 6.4, hybrid system designers claim that hybrids have been successful in reducing the fuel consumption at remotely located telecommunications sites in Antarctica, Canada, and the United States. The HR 3 hybrid power system consisting of an 84-kWh battery bank and a 1.2-kW solar array and operating on Calvert Island off the coast of British Columbia has cut diesel fuel consumption substantially. The system operator claimed that overall the hybrid system reduced fuel use nearly 90%—equivalent to half the maintenance cost of a conventional diesel engine. At Norway's Hamnjefell telecommunication station above the Arctic Circle, an HR 3 hybrid power system has met more than 70% of the site's electrical loads since 1985 [4].

6.8 Hybrid Systems for Village Electrification

Historians believe that 1.5 to 2.2 billion people have no access to electricity. Extending utility power service from cities to remote villages in developing countries presents a challenge to the city utilities. Construction of a coal-fired, nuclear, or hydroelectric power plant to electrify an individual village is both financially impossible and illogical. The most practical and economical answer is installing high-voltage transmission lines few miles away from villages with large populations, but this is not a sound scheme for bringing power to villages with populations of a few thousand. This approach is attractive for villages in undeveloped African and Asian countries. In developed countries such as the United States, Canada, and the European Community, electrification of villages and rural regions can be accomplished by distribution lines installed by commercial power companies.

6.8.1 Successful Examples

Since the cost of building conventional power plants is very high and villages contain no manufacturing facilities to purchase power, a stand-alone power system composed of a micro wind turbine and solar array is desirable for electrification. Batteries are not needed to meet emergency conditions because wind energy is available 24 hours a day and the solar energy is available for a minimum of 6 hours on a sunny day. This design configuration is cost-effective for village electrification and does not require scheduled maintenance. Storage batteries can be added if necessary.

A stand-alone power source consisting of a micro wind turbine and solar array operates successfully in the Mexican village of Xcalac on the Yucatan Peninsula. Bergey Wind Power installed the hybrid power system that utilizes both wind and solar energy sources. Bergey erected the first Mexican wind farm including a 60-kW array of six micro wind turbines, each 7 m in diameter, a 12-kW PV array tied to a large battery bank, and a 40-kW inverter. The complete power system cost about $500,000 which offset the construction of a proposed $3.2 million utility power line to this remote village.

Several hybrid power schemes successfully operate in Danish and Dutch coastal villages. In some third world countries, village electrification schemes are considered for pumping water, grinding grain, and handling other tasks that do not require utility-grade electricity, thereby eliminating the need for costly storage batteries. Batteries and inverters are needed only for loads that require constant frequency AC power.

6.8.2 Application of Village Electrification for Water Pumping

Most village farmers rely on irrigation water. The energy needed to pump water is a function of the volume and the height to which the water must be lifted. Specific details on pumping heads using wind turbine technology are shown in Figure 6.6. The figure clearly identifies the critical elements of a micro wind turbine: friction head, static water level, water outlet pipe, pumping head, total discharge head, and total dynamic head from a draw down reference line. The wind turbine takes kinetic energy available at the installation site and converts it into mechanical energy that drives the pumping motor coupled to an electrical generator. The amount of water pumped depends on the wind parameters and the various heads or heights involved.

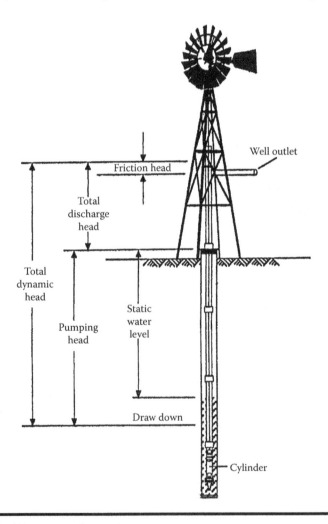

Figure 6.6 Water pumping system.

Table 6.5 Sample Calculation and Breakdown of Total Dynamic Head for Wind Turbine-Based Water Pumping System

Type of Head	Magnitude of Head (ft)
Static	100
Draw-down level	5
Discharge	10
Friction head for pipe	12
Friction head for 90-degree elbow	18
Total dynamic head	145

The static head is the most important parameter and is defined in a very practical way. For example, one plans to store water in a tank with an inlet pipe 5 ft above the ground up a small hill with a 5-ft gain in elevation or height. As to the pipe and storage tank, a 100-ft pipe (1-in. diameter) requires three 90-degree elbows to route the water into the storage tank. A breakdown of the total dynamic head for this system is summarized in Table 6.5. The numbers in the table are valid only for this example.

Unless the water storage tank is sited below the well outlet pipe as illustrated in Figure 6.6, a packer head or additional head is required to pump sufficient water into the tank. The packer head facilitates faster pumping above the height of the well into the storage tank. In the past, farm windmills provided domestic water and sometimes included storage tanks in the towers. Other windmill operators delivered water to a nearby tank on a raised platform of suitable height. Packer heads are required when an outlet is higher than a well opening.

6.8.3 Estimating Pumping Capacity of Farm Windmill

Early farm windmills directly coupled the wind turbine rotor via a crank to a sucker rod not clearly shown in Figure 6.6. The sucker rod lifts a column of water in the well as illustrated in the figure. Thus, a windmill lifts the sucker rod on every revolution of the turbine rotor. Under low wind environments, the weight of the water in the well would often stall the rotor, thereby bringing the system to a halt.

In recent years, most of the mechanical wind pumps have been back-geared to increase the mechanical power to be generated. Most current windmills use a transmission technique to enhance the mechanical advantage of a rotor under light wind conditions. Most back-geared windmills lift the sucker rod once every

● Well outlet pipe

● Wind turbine propeller

● Tower foundation

● Base cylinder

● Outlet control valve

● Discharge head monitor

Figure 6.7 Outlet components of pumping system.

three revolutions. This technique slightly increases the complexity of a windmill but enables it to pump more water under light winds. Most coastal countries deploy wind turbines to pump water for agricultural applications [1]. Pumping system components are shown in Figure 6.7.

6.8.4 Hybrid Wind Turbines for Economic Electrification

Many third world nations are scrambling to construct and install hybrid power systems to meet electricity demands in thinly populated regions. Energy planners for village electrification schemes have performed comprehensive trade-off studies and concluded that hybrid power systems that feature small wind turbines or micro wind turbines because of their relatively low procurement cost are best suited for village electrification schemes because they can supply electricity in a short time at minimum installation cost. In addition, as central power systems expand to villages, the hybrid power systems can be removed and sent to other remote villages where electricity is badly needed. According to energy planners, this scheme will allow electrification to expand rapidly with minimum capital investment.

A hybrid power system of a low-cost wind turbine of moderate capacity and a solar power system can provide nonstop power almost 24 hr a day. The benefits of the electricity generated by such hybrid systems include reasonable procurement cost, reliability, and maintenance-free, unmanned operation.

6.9 Multitasking Wind Turbines

Wind turbines designed and developed by Bergey Power Systems are best suited to perform several tasks simultaneously. Bergey wind turbines can charge the batteries needed for various applications and generate electricity compatible with the utility specifications established by the Public Utility Regulatory Policies Act of 1978. Furthermore, the Bergey Excel wind turbine is powerful enough to meet electricity requirements of an all-electric home.

Wind turbines from other sources are also capable of performing tasks simultaneously. Such wind turbines generally have output capacities to 1 kW with procurement costs exceeding $2500. Small villages in third world countries, particularly in Asia and Africa, cannot afford such systems unless the installation costs are paid by their central governments or through United Nations community development programs. Typically, 1 kW of electricity can provide energy for ten households in an underdeveloped village. Economic studies of energy planners indicate that a few 100-W capacity micro turbines could meet the energy requirements for a small village a minimum cost without compromising reliability and system performance. Sophisticated electronic controls and switches generally needed for 100-watt wind turbines are not required for micro turbines.

Wind turbine designers in Denmark decisively demonstrated that micro wind turbines or medium-capacity wind turbines have successfully powered homes, farms, and small businesses. In villages in Arctic regions and other or extremely cold areas, power generated by wind turbines can heat homes at minimum cost and without greenhouse effects. It is important to mention that successful harnessing of wind energy requires an operating site that provides sufficient wind speed that can be captured by a turbine. The author's studies of wind parameters indicate generally that a wind speed exceeding 30 mph is necessary to generate large amounts of electrical energy, which is possible with high-capacity wind turbines. The studies further indicate that a wind speed close to 6 mph is considered adequate for micro or small wind turbines that are best suited to charge batteries, pump water, heat homes, and generate electricity for domestic consumption.

6.9.1 Low-Power Turbine Applications

Low-power wind turbines can handle several applications that do not require storage functions. The smallest power system on the market, known as a micro wind turbine, provides an electric charge to the surface of a metallic pipe to counteract galvanic corrosion in highly reactive soils. Stored power is not needed during calm wind environments because corrosion is a very slow process; when the wind returns, the system again protects the exposed metal. Small and micro wind turbines output a few watts of power. The micro devices are widely used to charge automobile batteries, power security lighting at remote or unmanned facilities, and pump water where supplies are not readily available. These micro wind turbines involve minimum procurement and installation costs and do not require scheduled maintenance. Based on a current market survey, a micro wind turbine can be installed for less than $2000 and can function as a stand-alone power system at a rugged, remote, and unmanned operating site.

6.9.2 Design Requirements for Irrigation

Most stand-alone power systems deploy small wind turbines for irrigation applications. Since the wind is an intermittent source of energy and irrigation requires large amounts of water, storage is often impractical and wind turbines typically drive a well pump in conjunction with a conventional energy source. Coupling wind turbines with conventional energy sources is fairly simple with an electric pump, but not so easy with a diesel-driven pump. West Texas State University [1] developed an ingenious device for mechanical coupling the variable output of a wind turbine with an irrigation pump deploying an overrunning clutch mechanism device.

When the wind is strong and the wind turbine is producing he full output, the wind machine mechanically drives the well pump entirely on its own via the overrunning clutch. When the winds are weaker, the turbine assists in driving the water pump. In brief, the conventional power source operates the water pump alone and the overrunning clutch assures that a constant volume of water is pumped regardless of the wind speed. This system is widely used in countries that face shortages of water for irrigation and is best suited for stand-alone power systems. Typical daily pumping requirements in gallons per acre for various crops at a wind speed of 6 m/sec or 12 mph are summarized in Table 6.6. Note that water requirements vary based on soil characteristics and ground temperature.

6.9.3 Annual Energy Outputs of Wind Turbines

Estimation of annual energy output (AEO) follows these steps:

1. Calculate the wind power density (P_d) at the installation site and height where the wind turbine will operate (watts per square meter).

Table 6.6   Water Pumping Requirements for Common Crops

Irrigated Crop	Pumping Water Requirement (gal/acre)
Village farm	6400
Rice	10700
Cereal	4800
Sugar cane	6900
Cotton	5900
Lawn and garden	26100

2. Find the swept area (A) based on a conventional turbine or HAWT rotor radius (square meters).
3. Use 8760 (24 hr × 365 days) as the number of hours per year.
4. Assume a reasonable value (percent) for the overall conversion efficiency (η) of the system.

Find the product of the first three parameters above; multiply the product by the assumed overall efficiency.

This means that the magnitude of AEO can be written as:

$$AEO = [\text{step 1}] [\text{step 2}] [\text{step 3}] [\text{overall efficiency of wind power system}] \quad (6.1)$$

$$AEO = [P_d \ A \ 8760 \ \eta] \quad (6.2)$$

6.9.3.1 Wind Power Density and AEO Computations

Wind power density is dependent on the average wind speed at a specified operating height. According to recent data collected by meteorologists, wind speed increases with installation height according to the 1/7 power law. Bergey Power Systems measured an average wind speed of 5 m/sec or 11 mph at a height of 10 m above the ground. For a 7-m diameter wind turbine installed at a hub height of 100 ft, the wind speed will be about $30.48 \times 10^{0.143}$ or $3.048^{0.143}$ times the initial wind speed—5.86, rounded up to approximately 6 m/sec. The power density at 6 m/sec wind speed is about 253 W/m². The swept area A can be written as:

$$A = [\pi \ \text{radius}^2] = [\pi \ 3.5^2] = [38.48] \quad (6.3)$$

Typical overall system efficiency can be expressed:

$$\eta = [\text{rotor efficiency}] [\text{transmission efficiency}] [\text{generator efficiency}]$$

$$[\text{power conditioning efficiency including yawing and wind gust effects}] \quad (6.4)$$

Assuming various efficiency values, the overall efficiency can be written:

$$\eta = [36\%] [85\%] [80\%] [80\%] = [20\%] \quad (6.5)$$

Inserting the values from Equations (6.2) through (6.4) into Equation (6.1) yields an AEO [253 × 38.48 × 20% × 8760] of 17,000 kWh per year. Note this value applies only to the above assumed efficiency values. Values of annual energy output levels as a function of wind speed, power density, total overall efficiency, and conventional rotor diameter are shown in Table 6.7. The estimated AEO values in

Table 6.7   Estimated Annual Energy Output from Wind Turbines (kWh/year × 1000)

Wind Speed (mph)	Power Density (W/m²)	Total Efficiency (%)	Rotor Diameter (m)			
			4	5	6	7
9.0	90	28	2.3	3.6	5.2	7.1
10.1	110	28	3.4	5.3	7.6	10
11.2	150	25	4.1	6.5	9.3	13
12.3	190	25	5.2	8.2	12	16
13.4	250	21	5.8	9.0	13	18
14.6	320	19	6.7	10	15	20

the table indicate that magnitudes of AEO increase with increases of wind speed, power density, and conventional rotor diameter. It is interesting that the overall system efficiency decreases with increases in wind speed arising from abrupt variations in speed and direction. Typical average wind speed at a 100-ft hub height is about 6 m/sec or 13 mph—approximately twice the wind speed in meters per second. The parameters in Table 6.7 provide useful information about the performance of micro and small hybrid systems well suited for stand-alone installations.

6.10 Summary

This chapter is dedicated to stand-alone power systems composed of small or micro wind turbines and solar arrays to minimize system procurement cost. Preliminary studies by the author indicate that stand-alone wind turbine systems plays a critical role where commercial power lines are not readily available. These systems deploy small or moderate capacity wind turbines and banks of batteries sized to provide electricity for household consumption during the winter season and emergency periods during summers. A back-up diesel-driven generator can charge batteries during extended calms or shortages of electricity. The studies further indicate that the high cost, poor reliability, and frequent maintenance requirements for early stand-alone power systems discouraged customers. However, recent research, design, and development activities have solved most of the shortcomings and people who live in remote locations and or lack access to commercial utility service now accept the stand-alone power systems.

The author's studies reveal that implementation of solar cell technology in conjunction with a small wind turbine at a remote site offers a cost-effective and reliable option for customers who are interested in stand-alone power systems. Use of a diesel-based system or battery bank as a stand-by power source combined with a small wind turbine in a remote location is not a cost-effective approach. In some remote locations where high levels of power are needed badly, a diesel-driven generator or high-capacity battery bank could serve as a back-up power source, but this approach is expensive and requires scheduled maintenance. Solar cell technology is cost-effective for stand-alone systems in remove locations that lack commercial power lines. Various design configurations of stand-alone systems with emphasis on reliability, safety, installation cost, and maintenance-free operation have been described.

Stand-alone power systems are suited for village electrification because of low installation cost and maintenance-free operation. A stand-alone power system consisting of a small or micro wind turbine and solar array can charge batteries in a remote location or small village, pump water for irrigation, and power mills for grinding grain [1]. Solar arrays can provide electricity for emergency lighting, security devices, and low-power appliances operating at 110 V and 60 Hz frequency in remote locations where utility grid lines are not available. When solar power is not available under cloudy conditions, a micro wind turbine with moderate capacity can provide the needed electrical energy at low cost and with minimum complexity. Recent surveys indicate that a micro wind turbine of moderate capacity can be installed for less than $2000. A micro wind turbine of lower capacity for individual residential use can be purchased for about $1500. A unique use of a micro wind turbine is applying an electric charge to the surface of a metallic pipe to counteract galvanic corrosion in high reactive soils. This protection is suitable for pipes carrying oil, gas, and other nontoxic liquids in remote areas.

This chapter discussed the hybrid design concept for a stand-alone power system and focused on costs and reliability. A hybrid stand-alone system may include a small fossil-fueled back-up generator to provide electrical power for lighting, security sensors, and electronic devices, but this combination requires trade-off studies to determine performance, reliability, and installation and operation costs.

Hybrid stand-alone systems (micro wind turbines and solar modules) are desirable for telecommunication installations, particularly, in remote and inaccessible areas. Reliable performance of remote telecommunications installations is vital, particularly during floods and after earthquakes and air accidents in jungles and high mountain terrains. The proposed hybrid stand-alone system with back-up power capability can provide continuous (day and night) electricity to a remote telecommunications installation. At some remote sites, wind speeds exceeding 9 mph are required to generate electricity from a stand-alone micro wind turbine system. A hybrid system combines two low-cost energy sources (a PV system and a micro wind turbine) that are capable of charging batteries at inaccessible locations.

Estimated values of annual energy output levels (1000 × kilowatt hours per year) for stand-alone or hybrid versions of micro wind turbines based on wind speed, overall efficiency, power density, and conventional rotor diameter are provided for system designers or customers interested in small stand-alone systems. Critical issues to consider before selecting a stand-alone power source include installation costs, maintenance-free operation, and reliable guarantees of continuous electricity. Advantages and disadvantages of such systems such as cost, reliability, and design simplicity are summarized. During the 1980s, thousands of small wind turbines with capacities ranging from 10 to 40 kW were installed in California. At the same time, the Danes were installing 50-kW wind turbines in coastal regions to power homes, farms and businesses, because they depend heavily on the generation of low-cost electricity by wind turbines.

References

[1] P. Gipe, *Wind Power for Homes and Businesses*, 1993, Chelsea Green, Post Mills, VT, p. 11.
[2] S. Mertens, *Wind Energy in Built Environments,* Multiscience, 1989, Essex, U.K., p. 29.
[3] J. Walker and N. Jenkins, *Wind Energy Technology,* 2002, John Wiley & Sons, Chichester, p. 18.
[4] M.O.L. Hansen, *Aerodynamics of Wind Turbines,* 2nd ed., 1992, James & James, London, p. 63.

Chapter 7

Wind Energy Conversion Techniques in Built Environments

7.1 Introduction

Installation of wind turbines in built environments presents complex problems that are rarely encountered in open environments. This chapter will provide detailed descriptions of the wind resources in built environments that can be converted into energy by deploying wind turbines. This conversion technique requires focus on maximum energy yield from a wind turbine. It deals with the integration of wind turbine and building aspects in such a way that the building structure concentrates the energy of the wind for turbine use only. Preliminary studies undertaken by the author identified three distinct principles of buildings that concentrate wind energy strictly on turbines. Wind energy concentrators fall into four categories described as follows:

- Wind turbines installed in ducts through adjacent buildings
- Wind turbines installed at the roof of a building
- Wind turbines installed at the sides of a building
- Wind turbines installed between two airfoil-shaped buildings

The aerodynamic aspects of the four distinct types of concentrators and their respective wind turbines will be investigated; integration techniques for achieving maximum energy yields of wind turbines will be covered. Aerodynamic investigations involve mathematical models, computer simulation of air flow, and verification of computed parameters with measured data obtained from an open jet wind tunnel if available. Computer simulation may be performed with a computational fluid dynamics (CFD) code if available. It is important to note that both verification techniques involving measurement and CFD calculation will be exploited to achieve the most accurate verification results.

This subsection provides broad descriptions of the critical design concepts of the energy yield of a wind turbine installation in a built environment. The chapter offers detailed descriptions of the average global wind speed in a built environment, wind speed in the vicinity of an installation, wind speed near buildings, and verification of the mathematical models of the three possible concentrator schemes cited above. This chapter also provides useful information on appropriate wind turbines for deployment in built environments. The advantages and disadvantages of three concentrator principles (including emphasis on reliable and cost-effective design) are identified. At-the-roof configurations and variations of installations in ducts through buildings in relation to overall turbine efficiency will also be explained.

7.2 Concentrator Configuration Requirements

This section describes the configuration requirements and performance capabilities of concentrator devices used by the wind turbines in built environments. It is critical that a concentrator configuration be optimum for a specific built environment. For example, if an at-the-roof configuration is considered, a sphere-like structure will able to concentrate three to four times the energy of an omnidirectional free-stream wind [1].

7.2.1 Sphere-Like Configuration

Preliminary studies performed by the author on sphere-like building structures revealed concentrator configurations that can overcome the problem of low average wind speed in built environments and provide energy yields comparable to those in rural areas where typical wind speeds exceed 15 mph.

7.2.2 In-Ducts-through-Buildings Configuration

The author's studies also show that in-ducts-through-buildings configurations involving two ducted ellipsoids in a cross with a duct at the cross center can concentrate the wind energy in an omnidirectional free stream wind by a factor of about1.5. The studies further stipulate that other concentrator configurations offer lower wind energy concentrator factors and hence this configuration will not be discussed further.

7.2.3 Close-to-a-Building Configuration

This configuration is very efficient in using acceleration afforded by a structure. It offers the most cost-effective solution compared to the other concentrator types.

7.2.4 Capabilities of Various Concentrator Schemes: Summary

Potential wind energy concentrator configurations in relation to the amount of wind energy converted in built environments and concentrator efficiencies have been evaluated. Cost estimates depend on the type of structure, its height, and characteristics of nearby structures. Brief studies concluded that the wind energy yields for all the possible concentrator configurations evaluated are limited because the wind turbines can only benefit from the concentrator effects when the devices are relatively small compared to the building size. Despite small size, concentrators deliver the wind energy where it is needed, to the built environment, and their energy yields represent savings on utility bills. This constitutes a better payback than would be available in a rural area. We can conclude from these statements that wind energy conversion in built environments via concentrator effects from buildings may be recognized as a renewable energy source.

7.3 Energy Design Buildings

Energy design building technique offers reductions in energy consumption through the design of a structure. Reduced zero energy design (ZED) and lower energy design (LED) building architectures are recognized as the most desirable renewable energy schemes. ZED and LED building owners are encouraged to use power from renewable energy sources, thereby reducing their dependence on utility companies. Energy planners believe that small renewable energy sources integrated into building designs will reduce consumption of energy from commercial utilities. This way the building owners can be reimbursed at a price set by the utility company, which is roughly three times the price paid to commercial utilities for renewable energy sources in rural areas. It is important to mention that the motives for integrating renewable energy sources into residential and commercial buildings are driven by environmental issues but a building must be designed to capture large amounts of wind energy for the wind turbines.

7.3.1 Requirements for Built Environments

Architecture requirements and turbine installation locations must be selected to produce renewable energy with simplicity and at minimum cost. Studies performed by aerodynamic scientists and engineers on built environment power

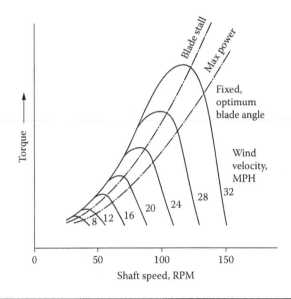

Figure 7.1   Requirements for renewable energy sources.

sources indicate that the earth's surface characteristics at a built environment are critical issues to consider in designing a renewable energy source in terms of high turbine efficiency, improved reliability, and torque versus shaft speed performance (Figure 7.1). The studies further indicate that a built environment generally has a higher roughness property than a rural area [1]. This high roughness produces a low wind speed (typically about 6 m/sec) compared to rural areas, where an average wind speed of 15 m/sec is not uncommon. Some local acceleration of the wind is required to achieve a viable wind energy yield in a built environment. Aerodynamic scientists claim that wind speed around taller buildings can be significantly higher than the average free stream wind speed in a built environment. It is possible to use this effect and design buildings with dedicated wind turbines that can utilize the increase of the average free stream wind speed.

Further improvement in the power output of a wind turbine operating in a built environment is possible by integrating concentrating devices in a design. Concentrating devices enhance turbine shaft performance as a function of wind speed. However, very little improvement in shaft performance can be achieved when the wind speed exceeds 16 m/sec as illustrated in Figure 7.2. Furthermore, concentrators add cost and excessive weight and reduce reliability. Based on comprehensive economic analyses, turbine designers conclude that a concentrator contributes a doubtful economic benefit. It is not sensible to build an expensive concentrator if a slightly bigger rotor would produce more power. Trade-off studies should be conducted before the deployment of concentrator devices to determine cost, reliability, and critical performance parameters. However, if a concentrator

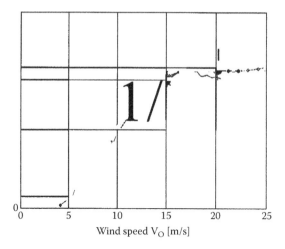

Figure 7.2 **Concentrator device.**

can be integrated into a wind turbine installation without excessive additional cost, the step should be taken.

Wind turbines capable of providing concentrated energy are known as building-augmented wind turbines (BAWTs). This section explains the aerodynamics of BAWTs, concentration schemes, and wind aspects in built environments. A comprehensive aerodynamics background is required to identify promising design configurations for efficient wind energy conversions in built environments. Successful installation of a BAWT power system requires comprehensive knowledge of wind energy, wind energy conversion, cost-effective concentration schemes, and built environments with high roughness factors. Brief studies by the author reveal that roughness factor depends on turbine operating height and operating environment. Variations in roughness factor are due to boundary layer profile changes. Variations in wind speed arise from changes in the operating height as illustrated in Figure 7.3. They decrease when vertical height approaches 100 m or so and variations thereafter are smaller than 0.06 m/sec around 180-m operating height.

7.3.2 Impact of Roughness Length on Wind Speed Parameters

Variations in roughness length are based on the type of terrain—an important parameter of wind speed and a function of boundary layer profile change; roughness varies with the terrain classification [3]. It is important to point out that computing variations in roughness length requires lengthy mathematical calculations involving critical meteorological data spanning 10 years or more. In addition, establishing the boundary conditions for wind energy conversion in a built environment is tedious and challenging. Analytical research tools include mathematical

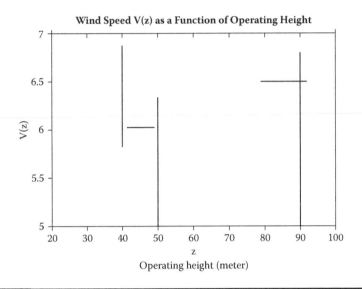

Figure 7.3   Effects of operating heights.

models of wind energy conversion in built environments. These models can be used to design the most cost-effective concentrator device for a particular wind turbine. After the boundary conditions (wind energy, wind energy conversion, and built environment) are defined, a designer can select a feasible and cost-effective option for a conversion in a built environment.

7.3.2.1 Estimating Wind Speed in City Environment

When the wind enters a city environment, the roughness height can be determined using the log law applicable outside the city boundary. The log law can also be used to calculate wind speed in a city environment. When air flow enters a city, it experiences a step-up in roughness height from Z_{01} to Z_{02}. The new boundary layer, known as the internal boundary layer, will develop a new roughness height designated Z_{02}. The effects of the step-up in roughness height do not occur instantaneously in the whole atmosphere and are limited to the height h_k of the internal boundary layer as shown in Figure 7.3. It is evident that the atmosphere outside the internal boundary layer behaves according to the upwind roughness Z_{01}. Under turbulence environments, the roughness height h_k will grow downwind of the roughness change, as shown in Figure 7.4.

An empirical model was developed to reflect the growth of the internal boundary layer based on dimensional analysis and bulk research data. This empirical model can be described by the following expression:

$$H_k(x) = [0.28 \, Z_{0,max}][x/Z_{0,max}]^{0.8} \tag{7.1}$$

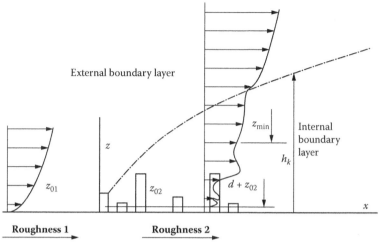

Symbols:

Z_{01} = Initial wind speed before entering the city

Z_{02} = Wind speed after leaving the city

h_k = Roughness height step of the internal boundary layer

Figure 7.4   Roughness factor changes.

where $Z_{0,max}$ is the higher of the Z_{01} and Z_{02} parameters. Equation (7.1) is valid for smooth-to-rough and rough-to-smooth variations in a wall region where $H_k(x)$ is less than 0.2δ and δ denotes a boundary layer height of about 1000 m for neutral air flow and strong winds.

It is important to mention that after some transient effects closely downwind of the step-up in roughness, the internal boundary layer will be logarithmic in a neutral atmosphere. Under these conditions, the log law for the internal boundary layer can be applied when parameter x varies between 500 and 5000 m. Scientists claim that within the approximation of the model of the internal and external boundary layers as shown in Figure 7.1, matching of the wind velocity above and below the parameter yields the velocity in the internal layer written as:

$$V_{int}(Z) = (1.32\ V_{ext})\ [\ln(h_k/Z_{01)}\ \ln(Z - d/Z_{02})]/[\ln(60/Z_{01})\ \ln(h_k - d/Z_{02})] \quad (7.2)$$

where V_{int} is the wind velocity in the internal boundary layer, V_{ext} is the wind velocity in the external boundary layer, d is the distance between the internal and external boundary layers, H_k is the roughness height or simply the height of internal boundary layer, Z_{01} is the roughness height of the internal boundary layer, Z_{02} is the roughness height of the external boundary layer, and Z is the roughness height of the internal boundary layer. Variations in roughness height from a step-up are shown in Figure 7.4. Experimental data of aerodynamic scientists indicate that a value of 60 m must be taken for parameter h_k because the roughness variations occur from high to low and vice versa so that their adverse effects cancel each other. The scientists further indicate that under turbulence conditions the roughness heights for the internal and external boundary layers are highly unpredictable and depend on the turbulence intensity (I_{turb}) which is written as:

$$I_{turb} = [\sigma_V / V_{ave}] \tag{7.3}$$

where V is the wind velocity, V_{ave} is the average wind velocity, and σ_V is the standard deviation of the wind speed V. For a neutral atmospheric condition, the standard deviation can be expressed as:

$$\sigma_v = [2.4\ V_0] \tag{7.4}$$

where V_0 is the wind velocity in the neutral atmosphere. Using above equations, the modified expression for the turbulence intensity is:

$$I = [\ln (Z - d/Z_0)]^{-1} \text{ if Z is greater than } Z_{min} \tag{7.5}$$

It is evident from Equation (7.5) that the turbulence intensity increases with increasing roughness and decreasing height to the surface of the earth.

7.3.3 Wind Potential in Built Environments

The energy available from the wind is proportional to the cube root of wind speed or velocity. Studies of aerodynamic engineers and meteorological scientists reveal that the highest wind speeds are generally found on hill tops, in coastal regions, and out at sea. Important parameters of wind include mean speed, direction, speed variations about the mean wind speed during short intervals, seasonal and annual variations, and variations resulting from height and temperature.

7.3.3.1 Atmospheric Boundary Layers

Temperatures in various atmospheric layers can affect wind speed and direction in each layer. At cooler temperatures, the air begins to sink in the lower layers of the atmosphere. Wind direction is affected by both the temperature and acceleration due to the rotation of the earth (Coriolis acceleration). The Coriolis acceleration [3] causes the wind flow from the equator to the poles to be deflected toward the east. The return flow toward the equator is deflected west and produces the trade winds.

7.3.3.2 Impact of Atmospheric Boundary Layer on Wind Speed and Variations

Wind flow is also affected by the local features such buildings, trees, and terrain features in the vicinity of the turbine installation site. Standing crops and water surfaces introduce fluctuations in air flow characteristics. In brief, the resulting changes in friction at the surface introduce fluctuations in wind speed and direction. Wind speed history can be measured by mounting a device called an anemometer on the nacelle of a wind turbine. Variations in wind speed can also arise from turbulence effects created by the turbine rotor and the nacelle housing.

The instantaneous wind speed (V_{ins}) can be defined as a mean wind speed (V_{mean}) plus the fluctuating component (V_{fluct}) and thus its expression can be written as:

$$V_{ins} = [V_{mean} + V_{fluct}] \tag{7.6}$$

The mean wind speed or velocity is typically determined as an average value measured over a 10-min span. The fluctuation of the wind speed is expressed in terms of the root mean square (RMS) value of the fluctuating velocity component and is also defined as the turbulence intensity. For very rough terrain including several trees and buildings, the turbulence intensity varies between 0.15 and 0.3 [3]. For smooth terrain surfaces, the intensity is typically 0.1. To date, the fluctuations in wind speed have been considered only in the horizontal direction.

7.3.3.3 Vertical Wind Speed Gradient

Research studies conducted by meteorological scientists indicate that the wind speed at the surface is zero due to the friction between the air and ground surface. The studies further indicate that wind speed increases with height above the ground, most rapidly near the ground, increasing less rapidly at increasing height as demonstrated by the values summarized in Table 7.1 that indicate that at a height about 2 km above the ground the change in wind speed approaches zero.

Vertical variations in wind speed or wind speed profiles at a particular installation site can be expressed by different functions. The two most common ones used to generate wind profiles include power exponent and logarithmic functions that were developed [3] to describe changes in mean wind speed as a function of height. The functions are based on experimental data collected by competent aerodynamic and meteorological scientists at various installation locations.

7.3.3.3.1 Power Exponent Function

This function can be written as

$$V(z) = [V_r (z/z_r)^\alpha] \tag{7.7}$$

Table 7.1   Wind Speed as Function of Wind Turbine Operating Height

Operating Height z (m)	Wind Speed at Operating Height V_z (m/sec)
40	22.97
50	23.49
60	23.92
70	24.29
80	24.62
90	24.91
100	25.18
125	257.4
150	262.20

where z is the height above the ground, z_r is the reference height V_r is the wind speed at the reference height, and α is the exponent representing a function of the roughness of the terrain. Calculated values of wind speed at the reference height are shown in Table 7.1. Wind speed parameters for calculating a wind profile as a function of surface roughness and exponent are summarized in Table 7.2.

These calculations use Equation (7.7) and assume 0.1 for the exponent and a wind speed of 20 m/sec at a reference height of 10 m. Close examination of these calculations indicates that at lower heights the variations in the wind speed decrease as the height increases. However, the variations in wind speed increase as the operating height exceeds 100 m.

Table 7.2   Wind Speed Parameters for Calculating Vertical Profile

Terrain	Roughness Class	Roughness Length z_0 (M)	Exponent α
Water surface	0	0.001	0.01
Open country	1	0.12	0.12
Farmland with buildings	2	0.05	0.16
Farmland with many trees	3	0.31	0.28

7.3.3.3.2 Logarithmic Function for Calculating Wind Speed Variations

Variations of wind speed as a function of operating height can be computed using the following logarithmic expression:

$$[V (z)/V (10)] = [\ln (z/z_0)/\ln (10/z_0)] \qquad (7.8)$$

where V (z) is the wind speed at operating height z (m/sec), z_0 is the roughness length, and V (10) is the wind speed (m/sec) at a reference height 10 m from the ground. Equation (7.8) can be used to compute the vertical profile of the wind. A wind speed of 5 m/sec is assumed at a reference height of 10 m for an open country terrain with a roughness length (z_0) of 0.12. Roughness lengths and the exponent values for various terrain types are shown in Table 2. The computed values of wind speed as a function of operating height (z) for an open country terrain are summarized in Table 7.3. These wind speed values are valid in the boundary layer near the ground as mentioned earlier. The calculated values using Equation (7.87) are summarized in Table 7.3.

It is important to point out that both functions can be used to calculate the mean wind velocity or speed at a given height if the mean wind velocity is known at the reference height (z_r). Parameter z_0 represents the roughness length for the type of terrain involved. Wind flow in the boundary layer near the ground is illustrated in Figure 7.5.

It is evident from these calculations that the difference in wind speed from 20 to 30 m height is 0.46 m/sec; the difference from 50 to 60 m is 0.20 m/sec; the difference from 90 to 100 m is 0.12 m/sec; the difference from 120 to 130 m is 0.09 m/sec; the difference from 140 to 150 m is 0.11 m/sec, the difference from 160 to 170 m is 0.07 m/sec; and the difference from 190 to 200 m is 0.06 m/sec. These values clearly show that the difference between two consecutive heights continues to decrease till the height approaches 170 m from the ground. Beyond 170 m, the difference in wind speed remains constant close to 0.07 m/sec.

7.3.3.3.3 Wind Statistics for Built Environments

Accurate wind statistical data for a built environment is absolutely necessary if cost-effective performance is the principal requirement for a wind turbine in that environment. Such data can be obtained from a histogram and a Weibull function for the probability of a given wind speed measured in 1 m/sec "bins" as illustrated in Figure 7.6. For example, a bin can represent a wind speed between 4.5 and 5.5 m/sec. Close examination of the histogram in Figure 7.6 indicates that the probability of a wind speed between 4.5 and 5.5 m/sec is approximately 10.4%. It is important to mention that the wind speed statistics are derived from data accumulated over

Table 7.3 Computed Wind Speeds in Open Terrain as Function of Wind Turbine Operating Height

Operating Height z, (m)	Wind Speed (m/sec)
20	5.78
30	6.24
40	6.57
50	6.82
60	7.02
70	7.20
80	7.35
90	7.48
100	7.60
110	7.71
120	7.81
130	7.90
140	7.98
150	8.09
160	8.13
170	8.20
180	8.27
190	8.33
200	8.39

several years. The histogram indicates the probability if a wind speed is within the interval given by the widths of the consecutive columns. When the width of a column decreases, the histogram becomes a continuous function called the probability density function. It considers seasonal variations and year-by-year variations over the years covered by the statistics. The cumulative Weibull distribution yields the probability that the wind speed will exceed a specific value.

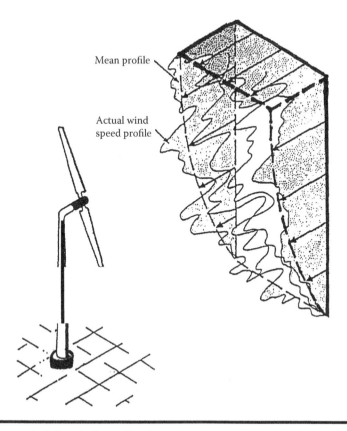

Mean profile

Actual wind
speed profile

Figure 7.5   Wind flow in boundary layer.

7.4 Local Wind Characteristics in Built Environments

The flow properties arising from building shapes and aerodynamic aspects of certain building types will be described. First, building shapes such as bluff and blunt types permit characterization of aerodynamic aspects. Second, air flow around buildings is characterized by the use of the Reynolds and Strouhal numbers. Finally, the probability distribution function of wind velocity at a specific location around a building should be derived.

7.4.1 Building Characterization

Brief studies performed by the author reveal that the first characterization of building shapes must distinguish aerodynamic buildings from bluff buildings. The studies further reveal that an aerodynamic building usually has a thin boundary layer attached to its entire surface. An aerodynamic building is characterized by a small wake. Bluff buildings are characterized by early separation of the boundary layer from the surface and a large wake. While bluff buildings exhibit phenomena that combine

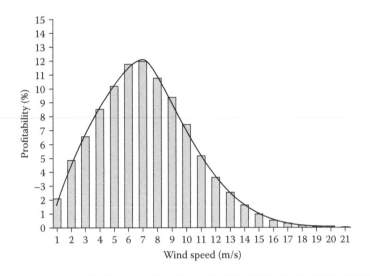

Figure 7.6   Histogram of wind speed.

flow characteristics of bluff and aerodynamic buildings, it is important to note that the characterization of a building as aerodynamic or bluff depends strictly on wind flow direction. Computer simulations by aerodynamic scientists clearly show the streamlines around an aerodynamic body and a bluff body in a parallel flow.

7.4.2 Streamlines around Buildings with Sharp Edges

Brief aerodynamic studies of buildings with sharp upwind edges reveal that the boundary layer separates at the upwind edges and separation bubbles form on the sides and at the top. The studies further reveal that the main stream is deflected around the building and a large wake downwind of the building is formed. Such phenomena typify a bluff building.

7.4.3 Drag Force Components

A building exerts two types of drag force components: pressure drag and viscous drag. The difference in local velocity around a structure results in pressure differences at its surface regardless of its dimensions. This essentially contributes to total or pressure drag. The no-slip condition at the surface of a building yields a contribution to total drag which is known as the viscous drag component of the drag force. Aerodynamic scientists believe that the pressure drag of an aerodynamic structure can be virtually determined by assuming potential flow around the building. Based on this assumption, the viscous drag component dominates the pressure drag component at an aerodynamic building. The viscous drag at a bluff building has the same magnitude as the viscous drag at an aerodynamic building but a different magnitude of the pressure drag due to the separated boundary layer. Consequently, the pressure drag component dominates the viscous drag component at a bluff building.

7.4.4 Air Flow Properties

Flow properties may be determined from the dimensions of the flow, the Reynolds number, and the Strouhal number. The flow can be two-dimensional (2-D) or three-dimensional (3D). Structures that have width and length are characterized as 2-D. Bodies comprising height, width, and length are categorized as 3-D. A building exhibiting a single very large dimension along with one or more smaller dimensions can be characterized as a 2-D body. The wind flow properties around 2-D (cylindrical) and 3-D (spherical) bodies differ, as illustrated in Figure 7.7. It is evident from this figure that a 2-D flow is approximated around bodies that have a single dimension ten times larger than another dimension. The flow around a 3-D object (sphere) will be relatively uniform and smooth.

The difference between the potential flow velocities at the sides of a sphere and a cylinder has critical implications for BAWT systems. These differences can be determined from a 3-D body by infinite elongation of the sphere in one direction. Bodies and objects have the same basic shapes; only their dimensions vary. The differences in flow around a sphere and a cylinder thus provide the differences in flow between 2-D and 3-D bodies or objects. Potential flow velocities at the sides of a cylinder and sphere are illustrated in Figure 7.7. The flow velocity at the surface of a cylinder or sphere can be determined using the potential theory. The flow velocity on the surface of a blunt body can be determined if the attached flow is assumed as a function of u_θ where u_θ is the flow velocity, θ is the angle measured from the axis of symmetry, and the wind is toward the sphere when θ equals 180 degrees.

Aerodynamic scientists believe that flow data obtained from simple inviscid hydrodynamics analyses may be replaced by analysis of the solid surfaces of flows

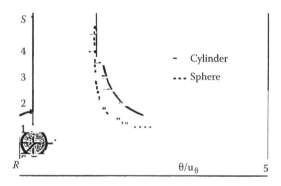

Figure 7.7   Wind flows around cylindrical and spherical bodies.

over spherical and cylindrical objects. Stream surfaces mat be replaced by solid surfaces so that the sphere flow stream surfaces represent the flow over a solitary spherical hill and cylindrical flow stream surfaces will correspond to flow over a long ridge. The hills do not have to be spherical or cylindrical and may be approximated by any stream surface.

7.4.4.1 Hydrodynamic Analysis of Flow over Sphere

For a flow over a sphere of radius a, the potential (φ) and stream (ψ) can be written [4]:

$$\varphi = [(Ua^3/2r^2) \cos\theta + Ur \cos\theta] \tag{7.9}$$

$$\psi = [(-Ua^3/Ur) \sin^2\theta + (Ur^2/2) \sin^2\theta)] \tag{7.10}$$

where r is the measured radial distance from the origin of the coordinate system which coincides with the center of the sphere of radius a, angle θ is measured from the axis of symmetry, and the wind flow is toward the sphere in the 180-degree direction. The tangential velocity v_θ and the radial velocity v_r components can be expressed:

$$v_\theta = (U \sin\theta) [1 + 0.5 (a/r)^3] \tag{7.11}$$

$$v_r = (-U \cos\theta) [1 - (a/r)^3] \tag{7.12}$$

Surfaces with constant flow speed $q = (v_\theta^2 + v_r^2)^{0.5}$ normalized by the flow at infinity U can be found from the solution of the following expression:

$$[(a/r)^{-6} + 4 (a/r)^{-3}\{(1 - 3 \cos^2\theta/1 + 3 \cos^2\theta)\}] = [4 (q/U)^2 - 1/1 + 3 \cos^2\theta] \tag{7.13}$$

Equation (7.13) is quadratic in the cube of (a/r), and can be solved using standard methods. Surfaces of constant dimensionless stream function (ψ/Ua^2) and constant speed (q/U) are represented in Figure 7.8 by symbols A and B, respectively. The surfaces shown in Figure 7.8 illustrate the limited disturbance of a flow field by the presence of a spherical object. Note that constant velocity surfaces are also known as isovelocity contours and are represented in Figure 7.8 by various values of the q/U ratio. It is important to point out that the flow at a 3-D bluff target or body can also move perpendicular to that plane with larger dimensions. Thus, it is logical to state that the flow velocities at the sides of a 2-D body will be higher. The boundary layer will separate at the along-wind sides of the blunt body and a wake will form, thus creating a possible difference between the potential flow resulting from the attached flow and the actual velocity at a Reynolds number exceeding unity.

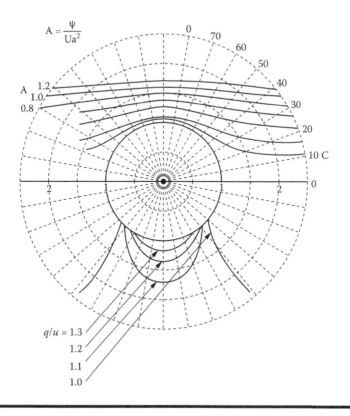

Figure 7.8 Surface functions.

7.4.4.2 Impact of Pressure Coefficient on Free Stream Quantities

The flow velocity at a Reynolds number much higher than unity can be computed from the measured data for the pressure coefficient (C_p), which is defined as:

$$C_p = [p - p_0/0.5 \; p \; u_0^2] \qquad (7.14)$$

where p is the pressure at the location where C_p is specified, u_0 is the free stream velocity, and p_0 is the pressure for the free stream. Using Bernoulli's equation, the tangential velocity component (u_θ) at a point where the pressure coefficient C_c exists can be determined:

$$u_\theta = [u_\theta (1 - C_p)^{0.5}] \qquad (7.15)$$

At the surface of a cylinder and sphere, the tangential velocity as a function of angle θ can be determined from a potential theory with an attached flow, which could be compared with a measured value at the pressure coefficient C_p at a separated flow according to aerodynamic scientists. Under these flow conditions, the Reynolds number varies between 200,000 and 400,000. The scientists further believe that the Reynolds number for a sphere and cylinder with separated flow can be accurate with error rate below 10% up to a circumferential angle θ equal to 90 degrees. When the ratio (r/R) is 1.1 and the angle is 90 degrees, the tangential velocity component at a radial distance can be given as:

$$u_\theta (r) = [0.9 \ u_\theta (\theta) \tag{7.15 a}$$

Four important conclusions may be drawn from the above data:

1. The tangential velocity at distance 0.1R from the surface of a sphere or a cylinder is reduced to 90% of the surface velocity.
2. The tangential velocity at the surface of a 2-D blunt body will be higher than the velocity at the surface of a 3-D blunt body.
3. The acceleration is independent of the absolute building size.
4. For a BAWT building, configurations that cause 2-D flow should be preferred above those that cause 3-D flow because the wind velocity near a 2-D building is considerably higher.

7.4.4.3 Theoretical Aspects of Reynolds and Strouhal Numbers

Basic knowledge of Reynolds and Strouhal numbers is necessary to visualize the flow patterns, particularly for BAWT) operations. The Reynolds number for a fluid can be expressed as:

$$Re = [\rho \ u_c \ D_c / \eta] \tag{7.16}$$

where ρ is the fluid density, u_c is the characteristic flow velocity, η is the dynamic viscosity of the fluid, and D_c is the characteristic dimension of a body completely immersed in the fluid. The Reynolds number can be defined as the ratio of inertial and viscous forces in the fluid flow. A high Reynolds number indicates that the inertial forces dominate the viscous forces; a low one shows that the viscous forces dominate the inertial forces. Therefore, a Reynolds number indicates whether the viscous effects in a fluid flow may be neglected.

Brief studies of Reynolds numbers and their effects seem to indicate that at certain Reynolds numbers, a conventional flow pattern can be observed in the wake of a bluff body. In addition, vortices of opposite signs are shed from the upwind edges of the bluff object in an alternate and regular way. This flow pattern is generally known as a vortex street. The vortices are very important for the building and BAWT installation. Because of their high local velocity components, these vortices induce a regular suction at the sides of a bluff object, thereby creating a resonance at the eigenfrequency of the bluff body. The magnitude of this eigenfrequency (f_e) can be determined from the following empirical formula:

$$f_e = [46/h] \text{ Hz} \qquad (7.17)$$

where h is the building height in meters. It is important to note that the regular high flow velocities at the sides of a bluff body can produce alternating power in a BAWT installation.

In a nondimensional form, the S (Strouhal) number represents the frequency of the vortices leaving one side of the building. For a 2-D building with a blunt or bluff shape, the S number [4] can be written approximately as:

$$S = [(0.21) \, C_d^{-0.75}] \qquad (7.18)$$

where C_d is the drag coefficient. However, by strict definition, the Strouhal number may be determined from the following equation:

$$S = [f_s \, D_c / u_0] \qquad (7.19)$$

where f_s is the shedding frequency produced by the vortices, D_c is the characteristic dimension or size of the body or building, and u_0 is the free stream velocity. From Equations (7.18) and (7.19), the expression for the shedding frequency produced by the vortices can be written as:

$$f_s = (0.21) \, [u_0 \, C_d^{-075} / D_c] \qquad (7.20)$$

If one allows the quasi-steady approach of the BAWT in the shedding region, a steady operation of a BAWT system can be achieved and a turbine with diameter (D_t) is installed if:

$$[1/f_s] > [6 \, D_t / u_0] \qquad (7.21)$$

The inverse of Equation (7.20) can be written:

$$[1/f_s] = [4.76 \, (D_c / u_0) \, C_d^{0.75}] \qquad (7.22)$$

Rearranging Equations (7.21) and (7.22) yields:

$$[D_t / D_c] < [0.8] \, C_d^{0.75} \qquad (7.23)$$

It is evident from Equation (7.23) that the ratio (turbine diameter to characteristic size of the building) must be less than 80% of $C_d^{0.75}$ where Cd is the drag coefficient that may vary for blunt to bluff bodies from 0.5 to 2.0. Research by BAWT engineers and aerodynamic scientists indicates that a D_d/D_c ratio below 0.5 will yield a stable BAWT operation. In other words, a ratio above 0.5 must be avoided if stable operation of a BAWT system is the principal requirement. In such cases, the BAWT diameter D_t is limited to D_t/D_c less than 0.1 to achieve higher system efficiency from the acceleration effects close to structures.

It is important to mention that the vortex street originates from the interaction of the flow on two sides of a building and is therefore more frequently observed behind 2-D buildings compared to 3-D buildings. Any structural adjustments that reduce interaction of the vortices may prevent the formation of the vortex street. Such adjustments are generally notched at upwind edges of the building or toward the apex. This BAWT installation is most common in the Netherlands.

7.4.3.4 Probability Function of Wind Speed for BAWT Installations

This subsection discusses the probability distribution of wind in the atmospheric layer at a specific BAWT installation site. Effects of various parameters on the probability of distribution at the location where the wind speed is changed by a concentrator structure will be identified. This subsection will also show the derivation of the energy yield of a BAWT installation from this probability distribution. It is important to note that the probability of distribution of the wind in the atmospheric boundary layer can be defined by a Weibull distribution, although the Weibull distribution works beautifully for wind speeds ranging from 4 to 16 m/sec at which the turbines operate most of the time. Variations in wind speed occur due to:

■ Location near a sharp-edged building
■ Location in a duct through a sharp-edged building
■ Entry of wind between airfoil-shaped buildings

A change of the free stream wind speed in the local wind speed near a building is Reynolds-independent. Therefore, a change in free stream wind speed caused by a major building used as a concentrator is Reynolds independent. The input wind speed u_i at the concentrator for the free stream wind from the windrose sector may be determined from the free stream wind speed $u_{0,i}$ at a certain reference height using the following expression:

$$U_i = [C_{r,,I} \, u_{0,I}] \qquad (7.24)$$

where the parameter $C_r i$ yields the Reynolds-independent change of the wind speed from $u_{o,I}$ to u_i. Some of these parameters vary based on height and surface roughness. However, the impact of installation height is more pronounced. For specific details on parameters involved in the equation for Weibull distribution, refer to standard texts on aerodynamics or hydrodynamics.

7.5 Impact of Built Environment on BAWT Performance

Adverse effects of a built environment on critical components of the wind turbine will be described along with details related to efficiency, reliability and safety. It is important to point out that built environments generate low wind velocity conditions and wind energy conversion is feasible at higher wind speeds. As stated earlier, higher energy is produced only by high wind speeds. Therefore the combination of wind energy conversion and built environment is not so logical from an energy yield view. Furthermore, harnessing of large amounts of wind energy requires site height capable of producing high wind speed conditions. In brief, installation sites well above the average building height are considered most promising.

7.5.1 Aerodynamic Noise Levels

Aerodynamic noise generated by wind turbine propellers is the most annoying factor for the people living nearby. According to aerodynamic scientists, the noise emission from a lift-driven horizontal axis wind turbine (HAWT) is approximately proportional to the fifth power of the tip speed. However, a low tip speed is highly desirable for a wind turbine in a built environment to keep aerodynamic noise to an acceptable level. In addition, all sources of vortex shedding from the turbine rotor must be avoided because the velocity differences and consequent pressure differences in the air flow are primarily responsible for aerodynamic noise generation. Wind turbine designers recommend that stall of the propeller blades must be avoided to minimize aerodynamic noise level.

7.5.2 Computation of Total Aerodynamic Noise at Installation Site

Local zoning laws require that total noise levels in residential areas must stay below a specified allowable maximum that is comfortable to residents. In addition, the allowable aerodynamic noise level (AENL) at nearby homes during the night is expected to be at the lowest level—approximately 40 dB (A) in the Netherlands. According to aerodynamic scientists, a noise level of 40 dB is also equal to the noise produced by a refrigerator at a distance of 1 m. They also note that this noise level equals the level in a rural area with a wind speed of about 9 m/sec or 18 mph. At a wind speed of 9 m/sec, very little energy can be generated by a wind turbine. Equations are available to compute noise levels. If a wind turbine is modeled as a

Table 7.4 Computed Values of Allowable Noise Level as Function of Distance R

Distance R (m)	Allowable Noise Level N_{an} (dB)
10	71
20	77
30	81
40	83
50	85
60	86
70	88
80	89
90	90
100	91

point source with spherical noise spreading, the noise at a certain distance R from the source can be computed. The reduction of noise (N_p) at location P can be calculated by the following:

$$N_p = [N_{an} - 10 \log_{10}(4\pi R^2)] \qquad (7.25)$$

where N_p (in decibels, dB) is the noise level at the source P, N_{an} (dB) is the allowable noise level at distance R (in meters) from the noise generating location. The computed values of allowable noise levels of a wind turbine in a built environment as a function of R to a sound pressure level of 40 dB are summarized in Table 7.4. Further calculations indicate that the allowable level varies, e.g., 97 dB at 200 m, 103 dB at 400 m, 106 db at 600 m, 109 dB at 800 m, and 111 dB at 1000 m.

7.5.3 Noise Arising from Nearby Wind Turbines

It is important to mention that the noise level from n identical sources ($N_{p,n}$) is characterized as the total sound pressure level from n noise sources operating at a specific point. The total sound pressure level from n wind turbines operating at the same location can be calculated from the following equation:

$$L_{p,n} = [10 \log_{10}(\Sigma 10^a)] \qquad (7.26)$$

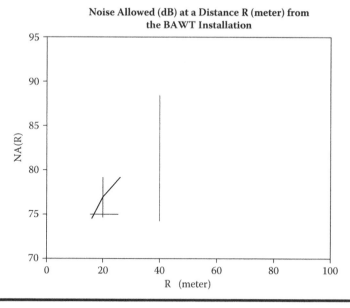

Figure 7.9   Allowable noise level.

where exponent α is equal to 0.1 $L_{p(j)}$, variable j varies from 1 to n, and the sum symbol has a bottom value of j = 1 and a top value of j = n.

Allowable noise level from wind turbines as a function of distance (R) can be estimated from Figure. 7.9.

Assuming a background sound pressure of 40 dB and ten wind turbines operating at the same location with a resulting sound pressure level at location P of 40 dB each, the total sound pressure level due to the background noise and ten wind turbines will be 50 dB.

The sound pressure level is an A weighted parameter denoted by dB (A) to identify the selective sensitivity of the human ear. This A weighting reduces the sound pressure with frequencies other than 1000 Hz because this particular frequency is the most sensitive of the human ear. Aerodynamic scientists claim that a wind turbine rotor generates a broadband aerodynamic noise with dominating frequencies of several kilohertz. Consequently, for the dominating frequencies of a BAWT installation or a wind turbine, A weighting does not have any impact on the noise levels generated by the turbine. Therefore the sound pressure levels in decibels around 1000 Hz frequency are approximately equal to the sound pressure levels in dB (A).

Aerodynamic scientists believe that two important issues are involved in noise emission. First, in contrast to low frequencies, the higher frequencies of several kilohertz seem to have a strong directional dependency. Furthermore, the higher frequencies may be heard only within the straight path of the source or as sound reflections from that path. The second issue identifies alternating noise levels as nuisances and scientists feel that these noise emissions should be ignored.

7.5.4 Induced Vibrations from Turbine Blades

Vibrations from wind turbine blades generally arise from pitch and yaw control mechanisms. Severe vibrations of blades occur during operation in gusty winds or mechanical turbulence. Amplitude and frequency of vibration depend somewhat on wind turbine configuration. For example, a HAWT system induces vibrations in the building at frequencies equal to the rotational frequency of the rotor ω_R and vibration frequencies equal to i $N_H \omega_H$ where i is the integer and N_H is the number of rotor blades.

The lowest induced vibration equal to ω_H is due strictly to mass imbalance and differences in aerodynamic loads of the blades. More blades on a rotor will produce low frequencies with varying amplitudes. The tower shadow and turbulence structures generate higher vibration frequencies. A shadow causes an induced vibration frequency equal to the product of rotational frequency of the HAWT rotor (ω_H) and the number of blades (N_H) at each revolution of the rotor. The rotor blades move through the low-velocity region upwind or downwind of the tower. Note that the blades that move through a mechanical turbulence structure with a velocity that differs from the average velocity will induce a frequency equal to N_H. This phenomenon is known as rotational sampling. If the turbine blades move through two regions with a velocity different from the average, they will induce a frequency equal to 2 $N_H \omega_H$ and so on. Thus, one finds an induced frequency caused by the rotational sampling equal to i $N_H \omega_H$ from HAWT systems.

The induced vibration frequencies from a VAWT are different from those described for HAWT systems. VAWT systems induce vibration frequencies in the building at the rotational frequency equal to ω_V due to mass imbalance balance and the difference in aerodynamic loads of the VAWT blades and frequencies equal to 2 i $N_V \omega_V$ caused by rotational sampling.

Based on these facts, the rotational sampling frequency of a VAWT is twice as high as that of a HAWT because the blades of a VAWT pass the mechanical turbulence structures twice—once at the upwind side of the VAWT and once at the downwind side. It is strongly recommended to avoid frequencies of a HAWT or VAWT close to the eigenfrequencies of the support structure including the building roof, walls, mast, and other components on which they are mounted. Any movement of the structure may produce resonance frequencies different from the blade vibration frequencies.

7.5.5 Shadow Flicker from Blades

Situations in which wind turbine blades are in the direct path between sun rays and the eyes of an observer and reflections of sun rays on wind turbine blades should be avoided. The problem of the reflections can be solved with dull paint. However, turbine blades located in the direct path of the sun's rays can be a nuisance to an observe near an operating wind turbine at visible frequencies below 20 Hz. Based on the preliminary studies performed by the author it can be stated that compared to the Darrieus, HAWTs have more problems in avoiding the low frequencies because of the single-blade passage where the Darrieus has a double-blade passage between the sunrays and the observer instead. Therefore, a HAWT will most likely cause serious hindrance because of the shadow flicking problem below 20 Hz frequency.

7.5.6 Turbulent Structures

Mechanical turbulence is strictly dependent on the average height or roughness of obstacles. This means that the average height H_{ave} of a building plays a key role in developing mechanical turbulence in a built environment. The largest turbulent structures in the wind can be scaled with this average characteristic size of the built environment. Aerodynamic scientists claim that large turbulent structures will change the local wind direction in the built environment in terms of the characteristic size of the turbulent structures (H_{ave}) and the average wind velocity (u_0). The time scale or duration of the mechanical turbulent (T_{turb}) structures can be found by dividing their characteristic size by the average wind velocity and writing the expression of the time scale as

$$T_{turb} = [H_{ave}/u_0] \qquad (7.27)$$

Computed values of turbulence duration as a function of turbulent structure size are summarized in Table 7.5. Note that a wind velocity of 6 m/sec is not so ideal from an energy yield view, while a duration of 5 sec for mechanical turbulence is highly undesirable from mechanical stability and structural integrity viewpoints . Brief studies performed by the author indicate that a minimum wind velocity of 9 m/sec is recommended to achieve high energy yield, while a maximum time scale or turbulence duration of 4 sec is highly recommended for mechanical stability and structural integrity. As mentioned earlier, wind energy increases at higher installation heights.

Table 7.5 Computed Values of Turbulent Duration or Time Scale

Turbulent Structure Size (m)	Wind Velocity(m/sec)			
	6	8	10	12
20	3.33	2.50	2.00	1.67
25	4.17	3.12	2.5	2.08
30	5.00	3.75	3.00	2.50
35	5.83	4.38	3.50	2.92

7.5.7 Impact of Stream Tube Length on Wind Energy Extraction

The momentum theory leads to the Lancaster-Betz limit that indicates a stream tube of finite length where the energy is extracted. Simulated studies of viscous flow by aerodynamic scientists indicate that a similar stream tube expands up to a certain diameter and then decreases in diameter due to reenergizing and speeding of the flow in and around the wake. The studies further indicate that the area where streamlines of the stream tube are roughly parallel is taken as the virtual "beginning or end" of the stream tube where the wind energy is extracted. Thus, for a viscous flow, the stream tube length is finite where the wind energy is extracted. The stream tube length is defined by the average path of particles of very small mass in the flow.

Computational fluid dynamics (CFD) calculations for a BAWT with an actuator show a virtual stream tube length approximately six times the diameter of the turbine rotor in the viscous flow. Based on this statement, the expression for the scale time T_s can be written as :

$$T_s = [6 \, D_{RV}/u_0] \tag{7.28}$$

where D_{RV} is the diameter of the rotor in the viscous flow and u_0 is wind velocity. This equation essentially characterizes the time needed to create a quasi-stationary stream tube for an actuator in which the Lancaster-Betz limit is applicable. Therefore, for a T_{urbo} less than $6 \, D_{RV}/u_0$), the varying effects become very important and the Lancaster-Betz limit based on the steady flow no longer applies. In this situation, to avoid the fast changes in load coupled with the varying effects, ensure that the T_{turb} must be greater than $6 \, D_{RV}/u_0$. Using Equations (7.27) and (7.28):

$$T_{turb} > T_s \text{ and } D_{RV} < [(1/6) \, H_{ave}] \tag{7.29}$$

where H_{ave} is the average height of the wind turbine or BAWT. Assuming an average height of 23 m, one finds a D_{RV} smaller than 4 m. This sample calculation provides an approximate estimate for the diameter of a wind turbine in a built environment. However, this sample calculation clearly shows that a wind turbine for a built environment is limited by rotor diameter. In other words, Equation (7.29) defines the critical rotor size or diameter for a wind turbine in a built environment. This rotor size limitation can provides an estimate of energy yield. Based on this conclusion, large amounts of electrical energy cannot be generated using wind turbine technology in built environments.

7.5.8 Yawed Flow Requirements in Built Environment

Three distinct boundary conditions must be satisfied for efficient wind energy conversion in a built environment: available wind energy, wind energy conversion technique, and built environment suitability. Based on these boundary conditions, the material in this section will formulate the point of justification in search of feasible and cost-effective options for wind energy conversion in built environments [4]. The author's preliminary studies reveal that three available wind turbine design configurations may be considered for a BAWT installation:

- Drag-driven
- Lift-driven
- Hybrid-driven

The rotor design configurations for these distinct wind turbines are illustrated in Figure 7.10.

7.5.8.1 Rotor Design Configurations for BAWT Systems

It is important to mention that for open rural areas, the rotor used for a HAWT presents certain advantages over other types. Wind turbine rotor design configuration for a built environment is subject to rotor size limits and restricted structural parameters. In other word, a BAWT system cannot use the same rotor used in a HAWT. The conversion of wind energy into mechanical power in the rotor axis may be realized with a drag-driven, lift-driven, or hybrid-driven rotor. Regardless of type, the efficiency or the power coefficient of the rotor is of critical importance. The expression for the power coefficient of a rotor can be written:

$$C_P = [P/(0.5) \, \rho \, A_{SA} \, u_0^3] \qquad (7.30)$$

where P is the power at the rotor axis, ρ is the air density, A_{SA} or A_t is the swept area of the rotor, and u_0 is the free stream wind velocity.

7.5.8.2 Drag-Driven Wind Turbine

The driving force of a drag-driven wind turbine rotor originates from the difference in drag of two rotating bodies as shown in Figure 7.10(A). The figure clearly illustrates the concept. The device consists of two cups. The cup with the spherical upwind side has the lowest drag. Consequently, this cup dissipates power as it moves in the direction of the incoming wind, while the other cup that moves downwind produces the power. Thus, it is characteristic for a drag-driven rotor that the power is produced by a body that moves downwind.

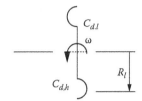

(a) Drag-driven wind turbine
using rotating bluff bodies

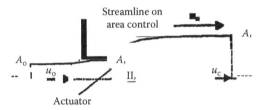

(b) Lift-driven wind turbine deploying an actuator

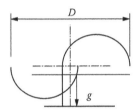

(c) Hybrid-driven wind turbine
using a Savonius rotor

Figure 7.10   Rotor configurations for drag-driven, lift-driven, and hybrid-driven wind turbines.

The torque (T) along the rotor axis is produced from the difference in the drag of the turning bluff bodies shown in Figure 7.10(A). It is evident from the schematic that the relative velocities on the turning bodies must be used to compute the drag. Based on the depicted rotor configuration, the torque of the rotor axis per square meter of swept area can be determined. The power generated on the rotor axis can be written:

$$P = (T \, \omega]$$
(7.31)

where ω is the rotational speed of the rotor and the torque (T) produced by the rotor is the product of drag coefficients of the rotating bluff bodies and the relative velocities multiplied by the air density (ρ). When the drag of the body that moves upwind is neglected ($C_{d,l} = 0$), the equation for the maximum power coefficient can be written as:

$$C_{P,max} = [(2/27) \, C_{d,h}] \tag{7.32}$$

where $C_{d,h}$ indicates the higher value of drag coefficient. The typical value of $C_{d,h}$ is 1.5. When this value is inserted in Equation (7.32), the maximum power coefficient for the drag-driven rotor is 3/27 OR 0.11. This means a drag-driven rotor converts a maximum 11% of the power available in the free stream wind into mechanical power at the rotor axis. If 90% efficiency is assumed for the generator and the gear box, a maximum of 10% of the electrical power is available from a drag-driven wind turbine rotor operating in a built environment—significantly lower than available power from a conventional wind turbine power source.

7.5.8.3 Lift-Driven Wind Turbine

Figure 7.10(b) is a diagram of a lift-driven wind turbine that deploys an actuator, which is defined as an energy-extracting device of small size dealing with a normal force on the surface that decelerates normal velocity through the disk. Note the wind flow is normal to the actuator plane. This is a working model for the rotor of a lift-driven wind turbine at infinite high value of parameter Å, which is defined as the ratio of the tip speed of the rotor and the free stream velocity ($\omega \, R_{at}/u_0$) where ω is the rotational speed of the tip of the rotor and u_0 the free stream velocity. Note the actuator is located in a stream tube in parallel flow with the entrance and exit of the stream tube at infinite distance from the actuator as illustrated.

The control area as shown in Figure 7.10(B) is bounded by the stream tube that forms the boundary of the flow through the actuator [5]. Hydrodynamic experts demonstrated through experiments that the pressure forces acting on the control area do not contribute to the momentum balance. Therefore, the conservation of axial momentum within the control area is assumed irrotational and the divergence-free flow shows that the thrust force acting on the actuator (F_a) can be written as:

$$F_a = \rho \, [u_0^2 \, A_0 - u_e^2 \, A_e] \tag{7.33}$$

where u_0 is the flow velocity at the input, u_e is the flow velocity at the exit, A_0 is the area of the flow stream at the input, A_e is the area of the flow stream at the exit, and ρ is the air density. Based on the conservation of mass theory within the stream tube:

[Mass of flow at inlet flow] = [mass of flow within tube] = [mass of flow at exit] (7.34)

Equation (7.34) can be rewritten as:

$$\rho[u_0 \, A_0] = \rho[u_t \, A_t] = \rho[u_e \, A_e] \tag{7.35}$$

where u denotes the flow velocity, A is the area of the flow, and ρ is the density of air. The subscripts 0, t, and e indicate the input, tube region, and exit, respectively. From Equation (7.35), one can write:

$$F_t = [(\rho \, A_t \, u_t \, (u_0 - u_e)] \tag{7.36}$$

Using Equations (7.33) through (7.35) and Bernoulli's theorem on a streamline upwind and downwind of the actuator, an expression for the thrust force (F_t) acting on the actuator may be written as:

$$F_t = [(0.5) \, \rho \, A_t \, (u_0{}^2 - u_e{}^2)] \tag{7.37}$$

Dividing Equation (7.37) by Equation (7.36) yields:

$$U_t = [(0.5) \, (u_0 + u_e)] \tag{7.38}$$

Equation (7.38) is strictly based on the momentum theory that becomes invalid when the exit flow velocity is less than or equal to zero.

7.5.8.4 Deceleration of Air by Rotor

The deceleration of the air by a rotor can be expressed in terms of an induction factor. The velocity through the rotor is:

$$U_t = [u_0 - a] \tag{7.39}$$

where a is an induction factor. From Equations (7.38) and (7.39), the expression for the flow velocity at the exit becomes:

$$u_e = [u_0 \, (1 - 2a)] \tag{7.40}$$

This equation states that the induced velocity infinitely far downwind of the actuator (2 a u_0), is twice the induced velocity (a u_0) at the location of the actuator. This important conclusion that may be obtained via simplified vortex theory, is called bare actuator wake expansion by fluid dynamic scientists.

7.5.8.5 Power Absorbed by Actuator

The power absorbed by the actuator cane is determined by subtracting the output power level from the input power level entering the stream tube as shown in Figure 7.10. The expression for the power absorbed can be written as:

$$P_{asb} = (0.5\ \rho)\ [A_0\ u_0^3 - A_e\ u_e^3] \tag{7.41}$$

From Equation (7.35) for conservation of mass in a stream tube and Equation (7.41), the expression for the power coefficient (C_p) is:

$$C_P = (\ 0.5\ u_t\ \rho\ A_t)[u_0^2 - u_e^2]/[\ 0.5\ \rho\ A_t\ u_0^3] \tag{7.42}$$

Inserting Equation (7.38) into Equation (7.42), the expression for the power coefficient is reduced to:

$$C_P = [F_t\ u_t/(0.5\ \rho)\ A_t\ u_0^3] \tag{7.43}$$

Equating (7.30) to (7.43), the power absorbed by the actuator can be determined by multiplying the thrust force (F_t) by the flow velocity through the actuator This is another important result used by fluid dynamics engineers and scientists. Using Equation (7.38), the expression for the power coefficient can be rewritten as:

$$C_P = 0.5\ (1 - u_e/u_0)/(1 + u_e/u_0)^2 \tag{7.44}$$

The expression for the maximum power coefficient can be found by setting the derivative of Equation (7.44) with respect to variable u_e equal to zero. This operation yields:

$$u_{e,\ max} = [(1/3)\ u_0] \tag{7.45}$$

Substituting Equation (7.45) into the derivation of Equation (7.44) with respect to u_e:

$$CP,\ max = (0.5)[(u0 - 1/3\ u0)/u0]\ [(u0 + 1/3\ u0)/u0]2 = (0.5)[2/3]\ [4/3]2$$

$$= [1/3][16/9] = [16/27] = [0.59] \tag{7.46}$$

It is evident from this equation that a lift-driven rotor converts a maximum of 59% of the power available from the free stream wind into the mechanical power at the rotor axis—about six times more than the power available from a drag driven-rotor. The actuator device absorbs a maximum of 59% of the power available in free stream wind. The actuator concept offers a useful model for an idealized lift-driven rotor best suited for BAWT systems.

7.5.8.6 Hybrid-Driven Wind Turbines

As noted earlier, a Savonius rotor configuration offers the least compact architecture with reasonable efficiency. Therefore, a BAWT system with a twisted Savonius configuration has been considered for hybrid-driven application in built environments. Specific details of Savonius rotor architecture with a gap between the two turning bluff bodies are illustrated Figure 7.10(C). The top view of a Savonius [5] rotor shows a gap (g) between the two bluff bodies. The wind turbine acts like a drag-driven device. Due to the gap in the Savonius configuration, the bluff bodies are driven by drag and suction, which characterizes this configuration as hybrid-driven. The additional driving force is generated by suction, which yields a higher optimum power coefficient ($C_{P, max}$) for hybrid-driven rotor than for a purely drag-driven rotor. The highest optimum power coefficient has been reported for a two-bladed Savonius rotor with a gap to diameter ratio (g/D_t) ranging from 0.10 to 0.15. Laboratory measurements of a hybrid-driven rotor demonstrated a $C_{P, max}$ equal to 0.24—twice as good as a drag-driven rotor.

7.5.8.7 Comparison of Wind Turbines in Built Environments

The author's preliminary studies indicate that the projected area of a drag-driven rotor is close to the rotor swept area. In contrast, the projected blade area of a lift-driven rotor is typically a fraction of the area for a drag-driven rotor. It is evident that a drag-driven rotor requires a lot of material and also has the lowest power coefficient. A lift-driven rotor combines a high power coefficient with fewer material requirements. The studies further indicate that a lift-driven wind turbine is effective for delivering wind energy with simplicity and minimum cost in a built environment. Drag-driven wind turbines are therefore not in great demand.

Table 7.6 Summary of Boundary Conditions for Wind Turbines in Built Environments

Built Environment	Wind Turbine Element	Building
Wind	Rotor size	Acceleration
Noise emission	Yawed flow	Resonance
Vibration	Reynolds number effects	Structural aspects

7.5.8.8 Boundary Conditions for Wind Turbines in Built Environments

Table 7.6 summarizes details of the boundary conditions of wind energy conversion in built environments, with emphasis on built environment aspects, critical elements, and construction. Boundary conditions are critical because they provide the contours of wind energy conversion in built environments and the contours yield useful data for projecting BAWT performance.

7.6 Summary

This chapter is dedicated to the performance capabilities and limitations of BAWTs. Critical design aspects of wind energy concentrators that are required to boost wind velocity in built environments are summarized along with discussion about cost and complexity. Potential wind energy concentrator configurations such as sphere-like structures and in-duct-through-building configurations and enhancement of wind velocity are discussed. Requirements for building structures with BAWT systems are summarized. Variations of roughness length as a function of wind velocity are identified.

Characteristics of atmospheric boundary conditions are discussed in detail. Effects of built environments on wind speed and wind variations are identified. The significance of vertical wind speed gradient due to build environments is discussed. Wind speed parameters vital in calculating vertical wind profiles are summarized. Effects of sharp upwind building edges on wind velocity are discussed in detail. Important components such as pressure drag and viscous drag are briefly summarized. Eigenfrequencies of mechanical resonance due to bluff building structures are explained. Calculated values of allowable aerodynamic noise levels as a function of distance from the source are provided.

Effects of induced vibrations and shadow flicker of wind turbine blades are described. Wind speed parameters vital in estimating the vertical wind movements are summarized. Mechanical turbulence effects and their durations as a function of building height are discussed along with details about mechanical resonance frequency. Impact of free stream length on wind energy extraction is briefly discussed with emphasis on the advantages of the computational fluid dynamic (CFD) method. Performance capabilities and limitations of BAWTs (drag-driven, lift-driven, and hybrid-driven) best suited for built environments are summarized as are power coefficient (efficiency) factors. Mathematical expressions for power coefficients of BAWT systems are derived and the computed values of their power coefficients are provided. Important boundary conditions for BAWTs are identified with particular emphasis on wind energy conversion efficiency and cost-effective design aspects.

References

[1] S. Mertens, *Wind Engineering in Built Environments*, 1989, Multiscience, Essex, U.K. 2006, p. 15.

[2] P. Gipe, *Wind Power for Homes and Businesses*, 1987, Chelsea Green, Post Mills, VT, p. 35.

[3] J.F. Walker and N. Jenkins, *Wind Energy Techniques*, 2002, John Wiley & Sons, Chichester, p. 57.

[4] D.A. Spera, *Wind Turbine Technology*, American Society of Mechanical Engineers Press, New York, 1994, p. 17.

[5] A.J. Wortmann, *Introduction to Wind Turbine Energy*, Butterworth, Boston, 1983, p. 89.

Chapter 8

Environmental Issues and Economic Factors Affecting Wind Turbine Installation

8.1 Introduction

This chapter focuses on the environmental issues and economical aspects justifying wind turbine installations. The environmental matters such as physical restrictions, noise levels, tower design constraints, disturbances of local ecology, effects on radio communications and television reception, zoning restrictions, and impacts on bird life must be thoroughly evaluated prior to selection of an installation site. The economic aspects such as initial costs for analysis and design, fabrication, assembly and operational testing, transportation, final checkout testing, installation costs for turbine and tower, cost of generation of one kilowatt of electricity compared to electricity from other sources, miscellaneous costs for maintenance, replacement, and repair, and finally economic feasibility for installation at a particular site must be considered. It is important to note that feasibility is determined strictly on the bases of costs and the value of electrical energy generated based on preliminary estimates.

A customer may have several questions that require accurate answers from a wind turbine installer. Will a particular wind turbine be able to meet his electric consumption requirements? Will the selected system pay for itself and if so, within what time frame? Will the system be a sound investment? Is it worth all this trouble? The answers may be elusive. The correct and reliable answers to these questions are based on a number of speculative variables that are not subject to precise calculation, for example, inflation rate, interest rate, and the desired rate of return based on prevailing economical factors. In brief, there is no single straightforward way to analyze the economic issues surrounding installation of a wind turbine.

8.2 Environmental Factors and Other Critical Issues

Environmental factors and other critical issues must be seriously evaluated prior to selection of an installation site for generation of electricity from a wind turbine. It may be difficult or even impossible to obtain approvals from appropriate authorities to operate a wind turbine system in a restricted zoning area due to objections from residents in the vicinity of the installation site.

8.2.1 Choice of Installation Site

Installation site is the most critical choice. Utmost care should be exercised in selecting a site for a wind turbine installation. Significant amounts of shear and compression normally occur in a horizontal wind stream as it travels over the topographical contours and rough surface of the earth at any installation site. Meteorological data should be collected over several years to ensure that wind speeds of 20 to 30 mph are available at operating heights of 20 to 30 ft, where wind speed is normally measured by anemometers.

Shear generates lower wind speeds near the surface than at heights great enough for free wind flow to occur. Furthermore, the free flow velocity at heights far enough from the surface to be unaffected by surface shear is significantly larger than that of the winds at the surface or at anemometer heights of 20 to 20 ft where wind speed is normally measured. According to aerodynamics and fluid dynamic scientists, the wind speed near the surface of the earth increases close to the 1/7 power of the turbine height above the surface of the ground, over open water such as a lake, river, or sea, and over flat plains as illustrated in Figure 8.1. It is clear that wind speed (V) varies as $H^{0.4}$ due to tall buildings, as $H^{0.28}$ due to trees and homes, and from the surfaces of open water and plains, respectively. These wind speed-versus-height relationships are valid only for few countries where wind speed measurements have been made over 8- to 10-year spans. These values may not be valid in desert regions due to wide temperature variations within the 20 ft above the surface of the earth.

Wind measurements made over decades by Canadian weather bureaus at two installation sites in Southern Canada [1] at altitudes from 7 to 10 m exhibited slightly different power exponents. The extrapolated wind speed versus height relationship showed that the wind speed varied as 16 the power of height ($H^{0.16}$). It is interesting that the geographic distribution of average wind power density is

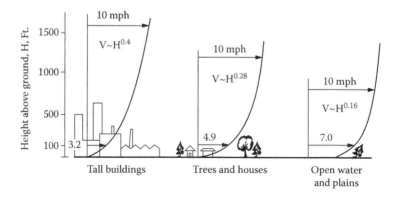

Figure 8.1   Wind speed dynamics.

even more skewed because it varies as the cube of wind speed, which is a function of installation height and surface conditions. Computed values of vertical wind speed (V) as a height (H) for various ground roughness levels are summarized in Table 8.1.

The computed values are for the vertical wind speeds that differ from the horizontal wind speeds that drive wind turbines to generate electricity. The horizontal component of the wind is important for turbine performance. A horizontal wind speed from 20 to 30 mph is ideal for most installations.

8.2.2 Wind Features and Their Effects

Brief studies performed by the author indicate that certain wind features produce certain effects such as maximum energy output and turbulence conditions

Table 8.1   Vertical Wind Speed as Function of Height for Various Ground Roughness Factors

Height H (ft)	$V = H^{0.4}$	$V = H^{0.28}$	$V = H^{0.16}$
100	6.31	3.63	2.09
500	12.00	5.70	2.70
1000	15.85	6.92	3.02
2000	20.91	8.40	3.37
3000	24.59	9.41	3.62
4000	27.59	10.20	3.77
5000	30.17	10.85	3.91

Table 8.2 Wind Features and Effects

Feature	Effect
Vertical wind speed variations	Even for level terrain, rough ground surface may result in turbulence conditions and at low-speed winds under a height of 30 ft; taller installation towers are required on rougher ground to reach levels at which laminar flow exists; laminar wind flow offers constant wind energy
Severe wind energy fluctuations	Wind energy by nature is unsteady and may be absent and thus require a storage scheme or back-up energy source for applications that require constant power
Omnidirectionality	Per HAWT theory, a turbine must always be vectored into the wind direction to achieve maximum energy output
Horizontal wind speed variation from location to location	Based on site surveys of meteorologists at various installations, optimum sites, particularly in contoured areas, are difficult to find
Nonuniform spatial conditions due to ground roughness and terrain variation	Wind turbine systems must be ruggedly built to operate with reliability in gusty winds and turbulence

that are detrimental to the stability and structural integrity of a wind turbine. Potential wind features and their effects are summarized in Table 8.2. It is critical that wind turbine customers understand the effects produced by various wind characteristics.

8.2.3 Terrain Modifications to Augment Wind Speeds for Improved Performance

Terrain modification may be required if the average wind speed at a specified installation location is below the minimum necessary to operate a wind turbine. Terrain modification may improve turbine performance with minimum cost and little complexity. A single type of terrain modification may not be suitable for all installation locations. Extreme care should be exercised in selecting a modification scheme to ensure highly reliable performance economically and simply. Potential terrain modifications are illustrated in Figure 8.2. Note how wind speed is augmented by terrain features such as hills, ridges, and well rounded hills. It is evident

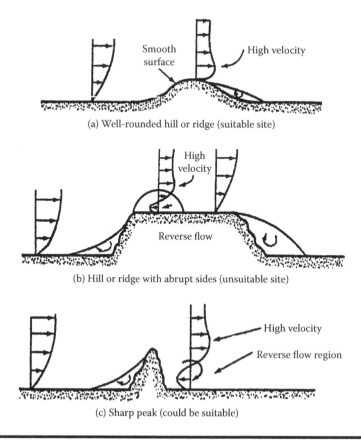

(a) Well-rounded hill or ridge (suitable site)

(b) Hill or ridge with abrupt sides (unsuitable site)

(c) Sharp peak (could be suitable)

Figure 8.2   Terrain modifications.

from Figure 8.2(a) that well rounded hill terrain offers high wind speeds and is thus well suited for wind turbine installations. Hilly terrain accelerates wind and is considered highly desirable from high wind energy yield view.

8.2.4 Impact of Volumetric Flow Rate on Performance

The author undertook aerodynamic of wind speed parameters that indicate that the volumetric flow rate of an air stream significantly impacts power output and power coefficient of a wind turbine. The studies further indicate that for a large blade area, a large reduction of windstream velocity will occur but the volumetric flow rate of the windstream will remain low. This means that any reduction in the volumetric flow rate of the windstream will affect the output power level and the power coefficient. These statements confirm that an optimum blade design maximizes the product of the volumetric flow rate and the pressure drop across the turbine.

8.2.5 Maximum Extractable Power

The maximum power that can be extracted from a wind turbine is the principal design parameter. The momentum theory states that the maximum amount of energy of a wind turbine that can be extracted depends on the volumetric flow rate of the airstream and that 8/9 of the kinetic energy of the wind stream is extracted while the airflow passes through the turbine. Under these conditions, the maximum loss in wind speed will be about 67 % of the initial velocity. Based on this principle, the expression for the maximum power that can be extracted from a windstream by a wind turbine can be written:

$$P_{max} = [0.295 \, A_T \, \rho \, V_0^3] \tag{8.1}$$

where A_T is the total blade area, ρ is the air density equal to 12.25 Kg/m³, and V_0 is the initial velocity of the windstream. Note the constant (0.295) includes the Betz coefficient equal to 0.59 that predicts the maximum fraction of power can be extracted by a wind turbine from a windstream. The equation of the power density can be written as:

$$P_{den} = [P_{max}/A_T] = [0.295 \, \rho \, V_0^3] \tag{8.2}$$

It is evident from Equation (8.2) that the power density increases by a factor of eight when the wind speed doubles as illustrated by Figure 8.3.

8.2.6 Power Coefficient

Power coefficient (P_C) is the second most important performance parameter of a wind turbine. Its magnitude depends on the ratio of the blade tip speed to free flow windstream speed. For propeller-drive wind turbines using three blades, the power coefficient attains the highest value of 0.59 when the ratio varies between 7 and 8. The power coefficient also depends on the rotor configuration and number of blades deployed. Typical power coefficients for a two-bladed wind turbine and various rotor configurations are summarized in Table 8.3.

For the Darrieus wind turbine rotor, the power coefficient attains a maximum of 0.35 at a ratio of peripheral speed to wind stream speed, not blade tip speed to wind stream speed ratio, because the peripheral speed of a Darrieus rotor is crucial. Rotor designs that are not listed in the table exhibit lower maximum coefficients at lower speed ratios than high-speed two-bladed horizontal axis rotors and Darrieus vertical axis rotors [2].

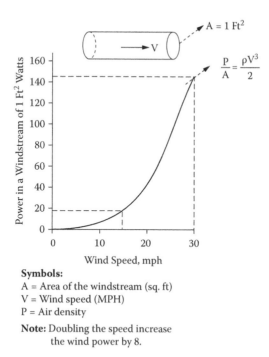

Symbols:
A = Area of the windstream (sq. ft)
V = Wind speed (MPH)
P = Air density

Note: Doubling the speed increase the wind power by 8.

Figure 8.3 **Power density as function of wind speed.**

8.2.7 Output Torque Performance

Aerodynamic design engineers believe that the output torque of a wind turbine is based on the rotor type and its parameters, rotor axis orientation, wind speed, and pitch angle adjustment capability. The torque developed by a horizontal axis rotor blade of fixed pitch angle varies with the wind velocity and the rotor shaft speed as shown in Figure 8.4. It is important to avoid blade stalling that occurs when the rotational speed of the blade is too slow in a windstream at a given velocity. Under these conditions, the output torque will decrease significantly and lead to unsafe stalling.

Table 8.3   **Typical Power Coefficients for Two-Bladed Turbines with Various Rotor Configurations**

Rotor Type	Blade Tip Speed-to-Windstream Speed Ratio	C_p
High efficiency	7 to 8	0.58
High speed	5 to 6	0.47
Darrieus	5.8 to 6.3	0.35

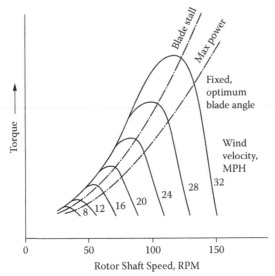

Wind speed (MPH)	Rotor shaft speed (RPM)	Maximum power occurs at	
		Shaft speed (RPM)	Wind speed (MPH)
8	32	50	13
16	62	100	26
24	83	125	29
32	126	200	34

Figure 8.4   Torque output.

If the rotational speed of a blade is too slow in a windstream of a given velocity, the blade will stall and thus reduce the output torque. Therefore, to achieve maximum output power or torque from a windstream with speed fluctuations, the pitch angle of the blade or the rotational speed of the blade must be adjusted. For this reason, most modern wind turbines are equipped with variable-pitch controlled blades. A controlled mechanism is required to adjust the pitch angles of the blades to maintain a constant rotational speed in the event of a change in wind speed or in output load of the turbine. It is clear from the curves in Figure 8.4 that maximum and maximum torque do not occur at the same wind velocity (mph) or rotor shaft speed (rpm) . Estimated values of torque developed as a function of wind velocity and rotor shaft speed are shown in Table 8.4. The maximum power levels as a function of rotor shaft speed and wind velocity are summarized in Table 8.5.

These tables illustrate the impacts of rotor shaft speed and wind velocity on the maximum torque and output power levels of modern wind turbines. Designers note that at wind speeds below rated values, the rotor speed must vary with the wind

**Table 8.4   Maximum Torque as Function of
Wind Velocity and Rotor Shaft Speed**

Wind Velocity (mph)	Rotor Shaft Speed (rpm)
8	32
12	41
16	62
20	72
24	83
28	97
32	124

**Table 8.5   Estimated Maximum Power Level as
Function of Rotor Shaft Speed and Wind
Velocity**

Rotor Shaft Speed (rpm)	Wind Velocity (mph)
50	13
75	19
100	26
125	29
150	35

speed to extract maximum power. They also indicate that this does not match the optimum operating conditions for synchronous or induction AC generators commonly used by high-power wind turbines. This mismatch poses a serious problem of interfacing a rotor with an electrical output generator.

The most practical approach is to allow the rotor speed to vary optimally with the wind speed and employ a variable speed, constant-frequency generator to achieve constant (60 Hz) electrical operation compatible with the commercial utility grid requirements. This way the power generated by wind turbines can be connected to commercial power grids. Suitable mechanical or electrical devices that can perform this task may be selected if cost and complexity factors do not undermine reliable system performance. Most wind turbine power systems deploy specifically chosen synchronous alternating current (AC) generators with fixed-ratio gear trains. This type of generator must be operated at constant input (fixed wind turbine rotor) speed to maintain synchronism with the electric current in the utility grid line to which it is tied.

8.2.8 Power Generated by Windstream as Function of Windstream Diameter

The output power (P) generated by a windstream relies on the wind speed (V) at the installation site and the wind stream diameter of the turbine (D)—the effective diameter of the rotor capable of capturing the wind energy. The output power generated by a wind turbine can be written as:

$$P = [(0.3927)\, \rho\, D^2\, V^3] \qquad (8.3)$$

where ρ indicates the air density (1.225 kg/m^3). It is evident from Equation (8.3) that wind turbine power output is proportional to the cube of the wind speed, but proportional to the square of the wind stream diameter. For example, doubling the wind speed can achieve four times more power output and doubling the wind stream diameter one can produce eight times more power. Computed values of power generated by a windstream as a function of wind stream diameter and wind speed are shown in Table 8.6.

Since most systems operate at wind speeds ranging from 20 to 40 mph as illustrated by Figure 8.5, no attempts were made to compute power output for wind speeds exceeding 40 mph. However, power generated by a windstream beyond speeds above 40 mph are shown in Figure 8.6. It interesting to note that considerable power can be generated by wind spread even at a wind speed of 50 mph with a wind stream diameter of 200 ft.

8.3 Problems Arising from Large Windstream Diameters

It is important to mention that increasing the effective diameter of a rotor will increase its cost, weight, tower structural complexity, and ambient noise level.

Table 8.6 Power (MW) Generated by Windstream as Function of Wind Speed and Windstream Diameter

Wind Speed V (mph/ft/sec)	Windstream Diameter D (ft)			
	50	100	150	200
10/15	0.0245	0.098	0.220	0.390
20/30	0.098	0.390	0.888	1.560
30/45	0.222	0.888	2.000	3.560
40	0.390	1.562	3.512	6.242

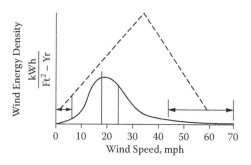

Figure 8.5   Wind speed variations below 40 mph.

Therefore, comprehensive trade-off studies should be performed to determine cost, reliability, and system complexity before considering significant increases in rotor size.

8.3.1 Higher Noise Levels

Noise generated by wind turbines is very annoying [3] and the decibel (dB) level ranges from the threshold of hearing to the threshold of pain. Scientists believe that doubling the power of a noise source by installing two wind turbines will increase the overall noise level by 3 dB—the smallest change most people can detect. It is

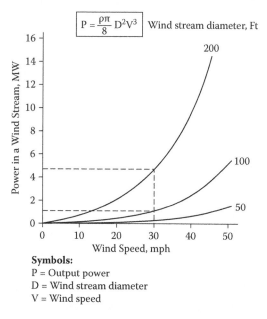

Figure 8.6   Wind speed variations above 40 mph.

important to mention that pain level is dependent on the pitch of the blades. The components of sound generated by a complex machine like a wind turbine are generated by many sources including the wind over the rotor blades and the whirring of the generator Furthermore, each sound component has a characteristic pitch, making it distinctive from the other sound components.

The noise measurements are dependent on the way the human ears perceive sound by using a scale weighted for the frequencies best heard by the human ear. Noise level is measured in decibels (dB)—the units commonly used to specify noise levels. The scale ignores sound frequencies we cannot hear. Note that impulsive sounds also elicit a greater response than sounds at a constant level. For example, kicking a garbage can draw someone's attention lot quicker than a person humming a tune. A wind turbine equipped with three blades will have a different sound profile than a two-blade turbine. Furthermore, a wind turbine with two blades spinning downwind of a tower will make a characteristic "whop-whop" as the rotor blades pass behind the tower. This impulsive sound may be missed by standard noise measuring equipment. It is interesting that many complaints about wind turbine noise in California have been directed at two-bladed downwind turbines. If a customer selects this type of wind turbine, he may need to consider the inherent effects from low-frequency sound components.

Another element is the time duration over which turbine noise can be heard. City or county noise ordinances generally specify a maximum level that must not be exceeded over a specified time frame. Some cities and counties weigh the time duration over which the noise occurs at various levels and frequencies. This complicates the task of estimating the impact from the noise of a wind turbine. Compared to noise levels from trains or airplanes that emit high levels infrequently throughout the day and night, a wind turbine may emit far less noise but does so continuously. Some may find this aspect of the wind turbine-based energy more annoying than noise generated by other energy sources. Noise acceptance is affected by subjective factors. If your community is unhappy about high utility rates, the sound from a wind turbine may add to community unhappiness.

Finally, the noise generated by a wind turbine must be placed within the context of noise levels from other sources. For example, if you live near an airport or a busy highway, a wind turbine will barely create a noise problem. Another example is noise from the wind. If an installation is in a high wind area, the wind turbine noise may not be bothersome because the ambient noise level of the windstream may impact the noise level generated by a turbine. It is important to distinguish the ambient background noise of an installation and the noise generated by the wind turbine. It should be the objective of the responsible authorities to limit increases in the total noise arising from a wind turbine installation in relation to noise generated by other sources [3].

8.3.2 Ambient Noise from Installation Site

Studies performed by wind turbine design engineers indicate that the ambient or background noise from the nearby trees varies from 51 to 55 dB (A) at a distance of 40 ft (12 m under a wind speed of 15 mph). Under these wind speed conditions, the noise from the nearby trees can mask the noise generated by a 10-kW wind turbine operating in the same wind environment. One must understand clearly the distinction between the background noise and the noise generated by a wind turbine. The background noise level is subject to surface conditions, while the noise generated by a wind turbine is based on the blade parameters and number of blades in the rotor. Finally, noise heard by nearby residents is the product of these two distinct noise levels.

Noise measurements made by aerodynamic engineers indicate that the ambient noise level from a 10-kW wind turbine is about 51 to 53 dB (A) at wind speeds of 25 mph (11 m/sec) [4]. The noise generated by the turbine varies from 54 to 55 dB (A) at a distance of 323 ft (100 m) to 53 to 54 dB (A) at 643 ft (200 m).

The noise level estimates predicted by the European Wind Energy Association for 300-kW wind turbines indicate that the noise from operation in a wind speed environment of 18 mph will drop to 45 dB (A) within 200 m. The overall noise level from 30 wind turbines of 10 kW each will be 45 dB (A) within 500 m. The noise estimates indicate that no wind turbine, no matter how quiet, can achieve levels better than the ambient noise. It is important to mention that the difference between the ambient noise level and the noise level generated by a wind turbine determines most people's responses. The main objective of the authorities should be limiting increases in the total noise to a level that should be acceptable to the residents in the vicinity of a wind turbine site.

8.3.2.1 Estimating Noise Levels from Wind Turbines

It is difficult but not impossible to estimate the noise generated by a wind turbine. However, an exact estimation is impossible because decibel level changes based on several unpredictable and fluctuating parameters. The following procedure is recommended to estimate the noise level from a wind turbine.

First, find the estimated level of noise emitted at some distance by a wind turbine at a specific installation site. Next, determine who will hear the noise and their distance from the site. It is also a good idea to estimate the sound pressure level those people will hear. It is extremely important to mention individual hearing sensitivities differ. This means that the estimation of sound pressure must be determined from the persons who have similar hearing sensitivities, which may not be possible in actual practice. Compare this result with any applicable county or city regulation and the ambient noise under conditions when the wind turbine will most likely operate, then compare the disposition of nearby residents toward your findings and anticipate the response expected from them.

8.3.2.2 Community Reaction to Wind Turbine Noise

Community acceptance of noise levels from a wind turbine site may lead to approval of the installation of additional wind turbines until the overall noise level exceeds the acceptable magnitude. A report prepared by Bergey Wind Power suggests that if there is any doubt whether a wind turbine installation will disturb anyone within a radius of 500 to 1000 m, it is wise to be a good neighbor and contact those who might be affected. Advise them of your installation plans and request their comments. Answer their questions as accurately as possible and make sincere attempts to incorporate their comments when designing your site. Bergey's report further indicates that the reaction of a community to the noise from a small wind turbine declines after the residents acclimate to the new sounds.

Since the early 1980s, wind turbine manufacturers have made significant efforts to reduce noise levels. They decreased aerodynamic noise levels regardless of turbine output capacity by sharpening the trailing edges of the rotor blades and using new tip shapes. Design engineers reduced the mechanical noise of larger systems by isolating the gearboxes from the nacelles and installing sound-deadening insulation. These techniques have made newer wind turbines significantly quieter.

8.3.3 Television and Radio Interference

Electrical design engineers believe that wind turbines deploying metallic rotor blades may disrupt radio and television reception. This is no longer a problem because modern blades are made from wood, plastic, or fiberglass. Of the 25,000 wind turbines operating in the United States and Europe, only a few have adversely affected communication signals. Electronic interference has been a problem only in remote areas where television and radio signals are extremely weak. Even in such areas, adverse effects have been rare and localized.

Small wind turbines are used worldwide to supply electricity for remote telecommunications stations for both commercial and military applications. Wind turbines would never have been selected to power such stations if the users saw any hint of electronic interference from the rotor blades. More than 5000 wind turbines operate close to the ridges of Tehachapi Pass in Southern California, but there has not been any interference with communication signals.

8.3.4 Quantitative Description of Wind Turbine Noise

Evaluating noise from wind turbines is very complex and the acceptable limits vary from country to country. For a comprehensive treatment of the subject, readers are advised to refer appropriate texts written by experts. Noise generated by a wind turbine can be quantified by measuring two distinct components: the sound power level component of the source (L_W) and the sound pressure level component (L_p) at the installation location. Both components are expressed in decibels (dB). The source power level component describes the power of the source noise; the sound pressure level component describes the noise at a remote point. Readers are advised to understand the clear distinctions of noise components. The sound power level component of a source such as a wind turbine can be defined as:

$$L_w = [10 \; Log \; (P_S/P_0)] \qquad (8.4)$$

where P_S is the sound power level of the source and P_0 is the reference sound power level (often 1 microwatt or 10^{-6} watt). The sound power level (L_p) varies with the wind speed at a turbine installation site. Calculated values of sound power level as a function of source power are summarized in Table 8.7. Most sound power levels have been measured between 96 and 108 dB (A) range at wind speeds close to 8 m/sec. The most common technique is to weight the measurements to reflect the sensitivity of the human ear. This is done with a weighting filter as illustrated in Figure 8.7(B). It is important to mention that the decibel (A) weighting function depends on noise frequency. The linear increase in weighting function continues up to 500 and attains its maximum value at about 2000 Hz and decreases thereafter. Wind turbine noise pattern as a function of distance from the installation is illustrated in Figure 8.7(A). The figure shows that wind turbine noise level decreases with distance from the turbine site.

Table 8.7 Computed Sound Power Level as Function of Wind Turbine Output

Wind Turbine Output (kW)	Sound Power Level A (dB)
1	90
10	100
20	103
30	105
40	106
50	107
100	110
200	113
300	115
400	116
500	117
1000	120

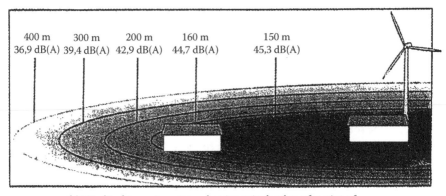

400 m	300 m	200 m	160 m	150 m
36,9 dB(A)	39,4 dB(A)	42,9 dB(A)	44,7 dB(A)	45,3 dB(A)

(a) Wind turbine noise pattern showing noise level as a function of
distance from the turbine site

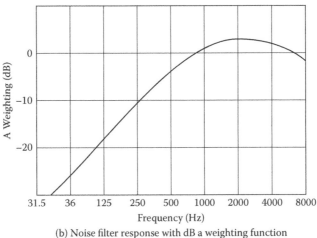

(b) Noise filter response with dB a weighting function

Figure 8.7   Weighting filter.

These calculated values are obtained using Equation (8.4) and assuming a wind speed of 8 m/sec. They will vary as a function of wind speed and operating conditions. In addition, the values may contain ±decibel (A) errors. The expression for sound pressure levels [5] can be written as:

$$L_p = [20 \, Log \, (p/p_0]$$ (8.5)

where p is the root mean square (RMS) value of the sound pressure level and p_0 is the reference sound pressure, which is often written as 20 µPa (micropascals) or 20 × 10^{-6} Pa (pascals). The pascal is the unit for expressing pressure. One pascal equals 6895 pounds per square inch (psi). In brief, the reference sound pressure equals

Table 8.8   Computed RMS Values of Sound Pressure Levels at 350 m

L_p (dB)	(p/p_0)	RMS p (lb/in²)
30	31.62	4.36
32	39.81	5.49
34	50.12	6.91
36	63.09	8.71
38	79.43	10.96
40	100	13.82
42	125.90	17.37
44	158.49	21.87
45	177.82	24.54

0.138 psi. Wind turbine engineers specify a typical sound pressure level ranging from 35 to 45 dB (A) at a distance of 350 m from a wind farm location. The (A) indicates that this parameter is measured using an A-weighted filter. As mentioned earlier, it is very common to weight the measurements to reflect the sensitivity of the human ear, which is done by applying a filter generally referred to as *dB A* or *dB (A)*. Inserting various values of sound pressure level (L_p) ranging from 35 to 45 dB into Equation (8.5), the RMS value of sound pressure can be calculated. Calculated values of RMS sound pressure levels at a distance of 350 m from a wind farm are summarized in Table 8.8.

Based on the table data, as L_p increases, the RMS of the sound pressure also increases. The operating range of 30 to 45 dB for parameter L_p is recommended for operating capacities in the range of 200 to 400 kW. For wind turbines with output power levels exceeding 1 MW, the sound pressure levels will be significantly higher. Persons planning to reside near high-capacity (2.5 MW) wind turbines are advised to live at least half a mile from the installation site.

8.3.4.1 Estimation of Noise Generated by Single, High-Capacity Wind Turbine

Sound power level data are based on wind turbine performance parameters and are generally available from the manufacturer. These data allow a user of a wind turbine to determine the sound pressure levels around the wind farm as a function of distance to ensure that the planning requirements are satisfied and no nuisance will affect nearby residents. Engineers recommend a simple model approved by the

International Energy Agency (1995) for determining the propagation of noise from a single turbine unit. The expression for this model can be written as:

$$L_p = [L_w - 10 \, \text{Log} \, (2\pi \, R^2) - \alpha \, R] \tag{8.6}$$

L_p is the sound pressure level at a distance R from the noise source with a sound power level of L_w and α is a frequency-dependent sound absorption coefficient. Both L_p and L_w are expressed in dB A or dB (A). Any noise emanating from a gear box can be easily detected by the human ear, according to designers. An additional noise level of 5 dB (A) is often added to the sound pressure figure for purposes of assessing acceptability. As mentioned earlier, a typical sound pressure level can be in the range of 35 to 45 dB (A) at a distance (R) of 350 m from a wind farm.

According to wind turbine designers, the product (α R) is small for high-capacity turbines operating at longer distances from the noise generating source compared to the difference between the parameters L_W and L_p. Under this assumption, Equation (8.6) can be rewritten as:

$$[L_W - L_p] = [10 \, \text{Log} \, (2\pi \, R^2] \tag{8.7}$$

Inserting several values of L_W and L_p in the recommended ranges and using Equation (8.7), we can verify the above assumption, namely, the product (αR) can be neglected for various distances from the noise generating source.

Sample Calculation:

L_W = 95 dB (assumed), L_p = 34 dB (assumed) and the difference ($L_W - L_p$) comes to 61 dB.

Inserting these values in Equation (8.7) yields

[61 dB] = [10 Log (2 π R^2)] or [6.1] = [Log (6.2832) R^2] or [Log (1258925.4)] = [Log (6.2832 R^2)] or [R^2] = [200169] yielding the distance R = 448 m.

Inserting various values of the parameters involved into the above equation, distances from a noise source can be calculated. The results obtained from Equation (8.7) are summarized in Table 8.9 and reveal that the larger the ($L_W - L_p$) difference in dB, the greater will be the distance at which the sound power level generated by a wind turbine will not disturb the nearby residents. In other words, the noise generated by a high-capacity wind turbine is acceptable to most people living at the distances specified in Table 8.9.

In summary, the noise from a wind turbine may be more obvious at low wind speeds than at high speeds because the general background noise caused by the

Table 8.9 Sound Power Level and Sound Pressure Level (L$_p$) as Function of Distance (R)

Sound Power Level L$_W$ (dB)	Sound Pressure Level L$_p$(dB)	(L$_W$ – L$_p$) dB	Distance from R (m)
95	45	50	126
97	45	52	159
100	45	55	224
108	50	58	317
98	39	59	355
95	35	60	399

wind trees is proportionally lower. The principal noise of wind turbines has two components: one is the aerodynamic noise from the rotor structure and the other the mechanical noise from the gear box, generator, and other structures. The mechanical noise has discrete tones, while the aerodynamic noise is usually broadband as illustrated in Figure 8.7(A). Aerodynamic noise is a function of vortices from the trailing edges and the tips of the rotor blades. A 300-kW modern wind turbine operating at a wind speed of 8 m/sec (or 15 mph) produces a noise level of 45 dB (A) 200 m from the installation site. The level reduces to about 37 dB (A) at 400 m distance. It is very important to install wind turbines at a minimum distance or more from nearby homes to ensure the noise level is acceptable.

8.4 Estimating Critical Performance Parameters Using Classical BEM Theory

Comprehensive aerodynamic studies by various scientists indicate that total wind energy production, mechanical shaft power, and other critical performance parameters of a wind turbine can be estimated using the classical blade element momentum (BEM) theory [6]. The calculated results are based on rotor radius, rotational speed of the turbine, hub height, wind speed, number of blades, and blade parameters such as angle of twist and chord size.

8.4.1 Mechanical Shaft Power

Mechanical shaft power generated by a wind turbine depends strictly on the wind parameters. Computed values of mechanical shaft power as a function of wind speed are shown in Table 8.10. It is evident from these data that a minimum mechanical shaft power close to 7% of the peak value occurs at a wind speed of 5 m/sec. The

**Table 8.10 Mechanical Shaft Power as Function
of Wind Speed**

Wind Speed (m/s)	Mechanical Shaft Power (kW)
5	42
10	318
15	516
20	562
25	571
30	512

optimum shaft power can be achieved over wind speeds ranging from 20 to 25 m/sec but shaft power continues to decrease as wind speed approaches about 23 m/sec. These results indicate that the BEM method captures the measurements very well at wind speeds below 25 m/sec as illustrated in Table 8.9.

8.4.2 Impact of Blade Design Parameters on Performance Level

The performance of a wind turbine is dependent on the blade parameters such as twist angle and airfoil characteristics (thickness and chord dimension)—both parameters are functions of rotor radius and typical values are summarized in Table 8.11. The typical values of twist angle and chord dimension shown in this table apply only to airfoils not thicker than 20% of the chord. Wind turbine blades can be designed to accommodate the specific airfoil best suited for a certain type of system. To maintain structural integrity of a wind turbine, an airfoil of about 40% of the chord thickness at the root of the blade must be used to absorb high bending moments. A thick airfoil is preferred to avoid adverse effects from centrifugal and Coriolis forces acting on the rotating blades. The rotational effects of the rotor, the blade geometry, and the airfoil characteristics affect the values shown in Table 8.5. It is interesting that both the twist angle and the chord dimension decrease as rotor radius is increased. Wind turbine efficiency achieves the maximum of 0.5 for a tip speed ratio between 9 and 10.

8.5 Justification of Wind Turbine Installation Based on Economics

Preliminary cost estimates for a commercial wind turbine installation must be prepared with a view to procurement cost for the turbine, tower, generator, and

Table 8.11 Typical Blade Parameters as Functions of Rotor Radius

Rotor Radius (m)	Angle of Twist (degr)	Chord Dimension (m)
4.5	20.02	1.627
5.5	16.30	1.597
6.5	13.08	1.540
7.5	10.06	1.481
8.5	7.45	1.420
9.5	5.86	1.355
10.5	4.85	1.294
15.5	1.358	0.956
20.5	0.018	0.262

associated electrical and mechanical components. Operating, maintenance, and site surveying costs including interest must be determined along with rate of return and break-even point. Exact estimates of wind turbine installation cost are difficult, if not impossible, because of the inherent problems associated with wind as an energy source.

8.5.1 Continuous Availability of Electrical Energy

It is important to mention that wind availability is not constant. If the wind slows down or stops, electrical power must be generated by other sources. This leads to a conclusion that it is necessary to have a back-up power producing system with the same capacity as the wind turbine if a continuous supply of electricity is vital. A moderate share of wind power in a system does not need a back-up capacity if the absence of a designated electrical load can be tolerated for a short time. For example, power companies in Sweden save water with hydropower dams when the wind blows and use this saved hydropower when the wind speed drops below the minimum since power consumption varies continually, particularly during the day regardless of season. In addition, every generating system has a built-in regulating capacity to adapt power generation as a function of wind speed variations and the output of the wind turbines. The electricity supply-and-demand proposition is critical role in determining the installed power capacity of wind turbines and estimating installation costs.

8.5.2 Estimating Procurement Costs

Procurement cost for a single wind turbine or a group of wind turbines will be very different from the cost for a wind farm utilizing 100 turbines. According to a survey on wind farms, the largest wind farm could include 100 wind turbines with an output power capability of 2.5 MW each. A survey of commercial systems [1] indicates that cost per kilowatt decreases for higher-capacity wind turbines; this may be verified from the cost data on propeller-based wind turbines summarized in Table 8.12. Preliminary cost surveys of megawatt class wind turbines indicate a 2.5-MW system made by the General Electric Company costs about $3.25 million. Procurement cost includes the wind turbine, accessories, electrical cables and wires, and tower structure.

8.5.3 Per-Kilowatt Electrical Energy Cost

It is difficult to determine the actual cost of electrical energy per kilowatt generated by a wind turbine. Calculations of the total cost of electricity produced must include the initial procurement costs of equipment, installation labor, and all other costs associated with operations and maintenance. Equipment procurement costs include the cost of wind turbine assembly, electrical generator, tower, and associated electrical and mechanical accessories. Methods of calculating costs range from very simple formulas to very large computer codes that generate large quantities of numbers and dazzling graphic displays. Furthermore, the true value of the extremely elaborate cost computations will remain questionable until sufficient data are accumulated to provide a solid and reliable basis for cost predictions.

Economic experts believe that one simple method is useful for rapid and reasonably accurate estimates. This method considers only the initial capital involved and

Table 8.12 Procurement Costs for Propeller-Based Wind Turbines

Manufacturer	Blade Diameter (ft)	Output (kW)	System Cost ($)
Aeolian Energy	32	10	17,750
Bergey	8.3	1	3,000
Dunlite	13	2	7,700
Enertech	0	4	16,000
Jacobs	23	10	20,000
Wind Works	33	9	29,500
Zephyr Dynamic	7.4	6.5	1,500

reasonable operating and maintenance costs. Wind turbine designers believe that it is customary to present the cost of a wind turbine in terms of cost per kilowatt of the nominal output power. It is evident from Table 8.12 that the procurement cost for a wind turbine decreases as the output power rating increases. That means cost per kilowatt is significantly lower for the megawatt class of wind turbines. It is important to mention that the specific power cost is a function of the prevailing wind speed at the installation site. This is not considered for the data shown in Table 8.12. However, data in the table assume a wind speed of 8 m/sec or about 16 mph.

8.5.4 Per-Kilowatt Electrical Energy Cost from Wind Turbine

As stated earlier, exact estimation of power cost per kilowatt because of unpredictable variations in wind speed, gear train loss, and the electrical generator type and efficiency. The actual power density of a synchronous AC generator driven by a gear train with a fixed gear ratio would be less than the density in the theoretical case. As noted in earlier chapters, power density of a wind turbine is strictly dependent on the cut-in wind speed or the velocity of the wind stream, rotor swept area, and the efficiency of the generator. Since the power density of a wind stream varies with both the velocity of the stream and the swept area of the rotor, the output generated by a wind turbine can be increased by locating a turbine in a region of higher wind speed and by increasing the diameter of the rotor. Both options need to be investigated with particular emphasis on procurement cost and installation complexity.

8.5.4.1 Empirical Method for Computing Annual Energy Output of Wind Turbine

Wind turbine engineers developed an empirical method to compute the annual energy output of a wind turbine based on the assumption that the specific power cost (dollars per kilowatt) is a function of the rated wind speed, but this factor is not included in annual energy output (E_O). The expression of the annual energy output of a wind turbine with a power rating of P_R can be written as:

$$E_O = [(P_R \, E_i) \, 8760] \tag{8.8}$$

where E_i is the energy integral (discussed in previous chapters in detail) and the 8760 (24 hours × 365 days) represents the hours per year. If the cost of a wind turbine is calculated on a per-kilowatt basis (C_{kW}) expressed as dollars per kilowatt, the annual cost is C_{annual}, and the annual maintenance cost is $C_{maintenance}$, a fraction of the initial cost, the total cost over N years of the wind turbine life will be approximately:

$$C_{total} = [C_{kW} \, P_R + C_{annual} \, N + C_{maintenance} \, N] \tag{8.9}$$

Due to inherent uncertainties in C_{kW} parameter, the total cost of the turbine for the life of N years can be written as:

$$C_{total} = (C_{kW})\ [(1/N) + C_{annual} + C_{maintenance}\ ]/[(8760)\ P_R]\qquad(8.10)$$

Sample Calculation:

Assuming the following values of the parameters involved, the total cost of the wind turbine per kW basis is:

$C_{annual} = 18\%$ (assumed)
$C_{maintenance} = 6\%$
Turbine life = 20 years (assumed)
Energy integral $E_i = 1.0$ (assumed)
$C_{total} = [11.9\ cents/kWh]$

The total cost of energy per kilowatt hour generated by a wind turbine is roughly three to four times higher than the cost generated by a coal-fired or nuclear power plant. Compared to a solar power generating source, the total cost of energy per kilowatt hour generated by a wind turbine is about 50% lower. The computed value of 11.9 cents/kWh has been verified by prominent wind turbine designers on the basis of a production run of 200 units with 180-ft rotor diameters and 200-kW ratings. Some turbine designers believe that a likely cost figure approaches 25 to 35 cents/kWh when realistic fabrication, maintenance, and tower costs are included. In summary, the calculated value of total cost of energy per kilowatt hour cannot be completely accurate because several assumptions are involved. If the values of the assumed parameters can be specified within an accuracy of less than ±2%, then the total cost of the energy per kilowatt hour can be specified with an accuracy of less than ±5%.

8.5.4.2 Cost of Electricity as Function of Turbine Cost and Life

Calculations of the cost of the electricity per kilowatt hour (cents per kilowatt hour) generated by a wind turbine as a function of turbine cost per kilowatt and useful life in years (N) indicate that the cost of the electricity generated per kilowatt hour decreases as the turbine cost per kilowatt decreases and the useful turbine life increases. The results obtained using the above equations and assumed parameters are summarized in Table 8.13.

High-capacity wind turbines with useful lives of 25 years are currently available. Systems of moderate and small capacities can attain operating lives close to 30 years. In order to further reduce the cost of electricity generated, one must select a

Table 8.13   Cost of Electricity (Cents per Kilowatt Hour) Generated as Function of Turbine Cost and Turbine Life

	Wind Turbine Cost (kW)		
Useful Turbine Life N (Yr)	*3600*	*3000*	*2400*
20	11.91	11.51	11.21
25	9.92	9.58	9.34
30	7.93	7.66	7.47

wind turbine with a per-kilowatt cost as low as $2000. However, a current market survey indicates that the cost is very close to $2500 per kilowatt. If the annual cost of money can be reduced below 15% and the useful turbine life can be increased to 30 years, the result can be a generating cost approaching 7 cents/kWh.

8.5.4.3 *More Elaborate Method of Computing Cost of Energy Generated by Wind Turbine*

A more elaborate expression for computing the cost of energy per kilowatt hour was proposed by various wind turbine designers [1] for high-capacity propeller-based systems. The cost of electricity generated by propeller-based wind turbines can be calculated by:

$$C_{kW} = [C_T f_T + C_L f_L + C_{OM} F_{OM} + C_O f_O]/E \qquad (8.11)$$

where C_T is the cost of the turbine, C_L is the land cost, C_{OM} represents operations and maintenance, C_O indicates the cost on a per-kilowatt basis, f is the multiplying factor, f_T is the turbine equipment fixed rate charge (18%), f_L is the land fixed rate (15%), f_R is the periodic levelizing factor [Equation (8.12)], and f_O is the annual levelizing factor.

$$F_R = [C_{CRF} (1 + e)/[1 + r]^{yains} \qquad (8.12)$$

where C_{CRF} is the capital recovery factor (8.9%), e is price the escalation rate (6 %), and yains indicates years after installation (typically 20). These parameters represent the recommended NASA accounting procedures for Mod-6H wind turbines and are considered adequate for all preliminary cost estimates pertaining to systems of moderate capacity.

8.5.4.4 Sample Calculation for Estimating Cost of Electricity Generated by Wind Turbines

The approximate cost of electricity generation by a wind turbine system can be determined from low-capacity systems (e.g., power rating of 10 kW, system cost of about $36,000 including installation, and assuming prevailing wind speeds exceeding 30 mph or 15 m/sec). If a wind turbine is located on 1 acre of land worth $1000 and the annual average wind speed is 22.5 mph*, the cost of electrical energy can be easily estimated. Assuming the value of the energy integral (R), the annual energy output from the 10 kW wind turbine can be given as:

$$E_{out} = [(10 \text{ kW}) (8760) \text{ (annual average wind speed/prevailing wind speed)}^3 (R)] \quad (8.13)$$

$$= [10 \times 8760 \times (22.5/30)^3 (0.8)] = 29,565] \text{kWh} \quad (8.14)$$

Assuming 18%, the annual cost of equipment, the actual equipment cost comes to

$$C_{ann.equip.} = [(18/100) (36,000)] = \$6,480 \quad (8.15)$$

Assuming that the annual replacement cost is only 5% of the original cost, in the third year the cost of replacement defined as the periodic levelizing factor (F_R) will be:

$$C_{Third.year} = [(5\%) (36,000) (F_R) \quad (8.16)$$

F_R is defined by Equation (8.12) in terms of capital recovery factor (8.9%), price escalation rate (6%), discount rate (8%), and years after installation (three assumed). This means:

$$F_R = [(0.089)(1 + 0.06)]/[1 + 0.08)^3] = [0.089 \times 1.06]/[1.08]^{20} = [0.0943/1.2597]$$

$$= [0.07486] \quad (8.17)$$

Inserting Equation (8.17) into Equation (8.16), the third year replacement cost is:

$$C_{Third.year} = [1800 \times 0.07486] = [\$134.75] = [\$135] \quad (8.18)$$

If 6% of the original cost of wind turbine ($36,000) is taken as the annual cost of operation and maintenance, the calculation is:

$$C_{opera + Maint} = [2 \times 0.06 \times 36,000] = [\$4,320] \quad (8.19)$$

---

* Annual U.S. wind speed is approximately 9.5 mph.

Using 15% as the land fixed rate charge, the cost of land equals:

$$C_{Land} = [15\% \times \$1000] = [\$150] \tag{8.20}$$

The total system cost is the sum of the results of Equations (8.15), (8.18), (8.19), and (8.20):

$$C_{Total\ system\ cost} = [\$6,480 + \$4320 + \$135 + \$150] = [\$11,085] \tag{8.21}$$

Based on dividing Equation (8.21) by Equation (8.14), the cost of electricity generated by this wind turbine is:

$$C_{elec} = [\$11,085/29,585kWh] = [37.47\ cents/kWh] \tag{8.22}$$

The cost of the electricity generated by a 10-kW wind turbine is about 37.5 cents/kWh—roughly four times the current rate for electricity produced by conventional power plants. It is important to point out that the electricity generated by a wind turbine located at a site with an annual average wind speed of 15 mph, will be about 20,762 kWh and the generating cost will increase to 53.4 cents/kWh.

8.6 Estimated Costs of Critical Components and Subsystems

We will now attempt to identify procurement costs for critical components and subsystems. The critical components include blades, rotor shaft, nacelle, gear box, generator, and pitch control unit. The tower, site foundation, and miscellaneous electrical and mechanical accessories are characterized as subsystem elements.

8.6.1 Cost Estimates for Critical Components

Procurement costs for critical components vary by manufacturer. If several suppliers can provide the same component, it is possible to negotiate a lower price. Fewer suppliers mean less room for negotiation and the possibility of paying the asking price. Generally, the costs of components of small commercially available wind turbines are lower than costs for high-capacity propeller-based systems. It is interesting that critical component costs for European wind turbines are slightly different from costs for U.S.-made systems. If a component for a European wind turbine is not available in the U.S., the result will be a higher price and longer delivery time.

It is difficult to determine the true procurement or replacement costs for critical components for wind turbines. European prices differ from U.S. prices because of stringent quality control procedures of U.S. manufacturers. In addition, procurement costs for critical components for augmented systems are much higher than those for standard wind turbines because of large stationary surfaces and expensive rotating elements. Due to the lack of reliable cost data and differences in estimating procedures, little has been done to achieve concrete and meaningful cost data covering critical components. Because of this lack, different cost estimating procedures must be used for large wind turbines with output power ratings exceeding 100 kW and small wind turbines rated at 1 to 10 kW. No attempt has been made to estimate component costs for augmented systems because of the complexity of components. Some procurement experts are reluctant to express their opinions about costs. The major issue in estimating component costs for large wind turbines may be the lack of agreement about the contributions of major components to the total cost of a system. The procurement costs for critical components of four large propeller based wind turbines are summarized in Table 8.14.

Based on the table data, the percentage rate of the total system cost exhibits enormous spread. The author used his engineering cost projection experience to prepare a medium cost estimate for the critical components of large conventional wind turbines. Note that the procurement costs for critical components used in European wind turbines manufactured in Denmark and Switzerland with power outputs from 1 to 10 kW are somewhat less expensive. Critical wind turbine components manufactured in the U.S. cost more because of higher wages and fringe benefits. Some turbine designers point out that for European wind turbines and the old U.S. Putnam machines, rotor costs vary from 10 to 20% of the 67% rotor cost contribution cited by Eldridge [5] for medium capacity systems. The rotor cost contribution varies from 67 to 40% for wind turbines with power ratings ranging

Table 8.14   Estimated Procurement Costs of Critical Components

Component	Percent of Total System Cost	Medium Percent Cost
Rotor blades	3 to 11.2	7.1
Gear box and generator	13.4 to 35.4	24.4
Hub, nacelle and shaft	5.3 to 31.5	18.4
Control system elements	4.2 to 10.2	7.2
Tower	5.3 to 31.1	18.2
Site and foundation	8.4 to 26.2	17.3
Miscellaneous engineering	3.2 to 11.4	7.3

from 100 to 1000 kW (1 MW). In brief, the large number of variables of wind turbine systems make it extremely difficult to determine the true costs of individual components.

8.6.2 Typical Design and Performance Characteristics of 2-MW Wind Turbine

Simply to provide general knowledge of design and performance parameters of high-capacity wind turbines, the author selected the 2-MW wind turbine designed, developed and manufactured by the General Electric Corporation. This company manufactures several high-capacity (2.5-MW) wind turbines and its units operate around the world. Major design and performance parameters of the 2-MW system [6] can be summarized as follows:

Rated power capacity = 2 MW
Rated and desired wind speed = 25 mph
Survival wind speed = 150 mph at rotor
Cut-in wind speed = 11 mph
Cut-out wind speed = 35 mph
Cone angle = 12 degrees
Inclination of axis = 0 degrees
Rotor speed = 35 rpm
Blade diameter = 200 ft
Blade twist angle = 11 degrees
Blade ground clearance = 40 ft
Operational life = 30 yr

8.7 Wind Turbine Towers

Tower structures are essential to provide the safe and reliable operation of turbines under a variety of wind conditions. During the 1930s, turbine towers in Great Britain were not very tall. The flat terrain and few obstructions at installation sites did not call for towers taller than 60 ft, despite the claim by aerodynamic engineers that "the higher the tower, the greater the power." By the late 1940s guyed tower heights ranged from 80 to 110 ft. As the technology matured, installers learned that cost-effective power generation and optimum wind turbine performance require tall towers. Designers were fully confident that wind speed capture and optimum power generation were possible only with greater tower heights. Wind turbine experts now understand power-robbing turbulence and its effects on performance. Manufacturers prefer taller towers because they want their products operate with utmost safety and optimum performance. In addition, the manufacturers believe that taller towers allow more design flexibility, ensure high wind energy capture, and are preferable if buildings and trees are present. By today's standards, a minimum tower height of 80 ft is recommended for moderate capacity wind turbines. When trees are nearby, a height of 100 to 120 ft (30 to 35 m) is recommended.

8.7.1 Tower Height Requirements

Height requirements vary by turbine rating. As stated earlier, the height requirements for moderate capacity turbines vary between 100 and 120 ft. Height requirements for micro turbines commonly used in residential settings are somewhat different. Micro systems generally deploy rotor diameters smaller than 3 m (10 ft) and are best suited for low-power applications. Because micro turbines are low-cost systems utilizing simple towers, the towers are not suited for turbines more than 60 ft above the ground. Tower heights for wind turbines with capacities exceeding 100 kW usually exceed 100 ft.

8.7.2 Mechanical Strength Requirements

A tower must withstand the forces and bending moments arising from high wind conditions and also provide the structural integrity to carry an expensive wind turbine and its accessories. In brief, the mechanical safety and integrity of an entire wind turbine system is depends strictly on the tower structure. Towers are rated by the thrust loads they can endure without buckling and bending. U.S. safety standards require manufacturers to design their wind turbine systems to withstand 120-mph (54 m/sec) winds without sustaining damage. The thrust on a tower structure at this wind speed is based on the rotor diameter of the turbine and its mode of operation under the defined prevailing conditions. Wind turbine manufacturers believe that a tower structure vibrating in resonance must be avoided because such vibration can exert tremendous force on the tower and lead to a catastrophic failure of the tower and turbine. In brief, creating loads on a tower structure must be avoided under all operating conditions.

8.7.3 Tower Classifications

Towers used for wind turbine installations fall into two major categories: free-standing (self-supporting) and guyed types. The Eiffel Tower is the best known example of a truss tower. Truss towers are generally more rigid than pole towers. A tubular or pole tower is another free-standing form produced in several different varieties. Towers can be designed to withstand any load condition. However, as the size of a wind turbine increases, the weight, cost ,and tower height increase proportionately. As the output capacity of a wind turbine increases, the components become bulkier and heavier, and more difficult and costly to move. In the U.S., truss towers are assembled in 30-ft sections. For small-capacity systems, the sections may be pre-assembled and welded together prior to delivery. For large-capacity turbines with outputs exceeding 100 kW, towers are shipped in parts and assembled at installation sites.

Installation of truss towers generally requires a crane. The tower is assembled on the ground at the site, then hoisted into place and bolted to the foundation using high-strength steel bolts. The author's studies of tower installation indicate that towers, particularly for medium- and high-capacity systems, must be assembled and bolted to tower foundations by experienced personnel as recommended by the turbine manufacturers.

Pole towers are best suited for low-capacity or micro turbines with output ratings below 10 kW and are generally fabricated from tapered steel tubes, inexpensive steel pipe, concrete, and even fiberglass. Both the tapered steel tubes and steel pipes can be guyed to provide additional strength. Steel pipes are widely used in guyed applications for small wind turbines (outputs below 5 kW). Pole tower installation generally requires a crane. For a micro turbine, a pole tower can be hinged at the base, tipped into the desired place with a gin pole, then bolted to the foundation.

In Denmark and Switzerland, manufacturers erect the hinged tubular towers with powerful hydraulic jacks. Pole and tubular towers present considerably higher erection costs than guyed towers but are only modestly more costly than truss towers. Free-standing pole and tube towers are more aesthetically pleasing than truss towers. In California, the tubular towers are visible at longer distances than lattice towers. According to wind turbine manufacturers, the lattice towers tend to blend into the background at distances above 2 mi (3 km). Pole towers require stronger foundations than truss towers to spread overturning force over a wider base. For small or micro wind turbines, guyed towers offer a good compromise of mechanical strength, erection cost, ease of installation, and appearance. However, they require more space than free-standing towers and are prone to crushing under heavy winds and turbulence conditions.

8.8 Summary

This chapter describes the environmental issues and critical economic aspects that justify the deployment of wind turbines to generate electricity. Effects of critical factors such as free wind flow, wind speed variation as a function of installation height, and ground terrain features on the power generation capability of wind systems are discussed in detail. Potential wind features and their effects, namely vertical wind speed, ground roughness coefficient, wind energy fluctuation, horizontal wind speed variation, nonuniform spatial conditions resulting in from ground roughness and terrain variation, and omnidirectionality are explained. Terrain modification techniques to enhance the volumetric flow rate and wind speed vital for improving performance at sites with marginal wind speeds are detailed.

Expressions for optimum power output and maximum power density as a function of wind speed and wind stream diameter are developed. Various methods for optimizing the output torque as a function of rotor shaft parameters, wind speed, and pitch-angle adjustment are outlined. Calculated noise levels from large wind turbines related to distance from the source are discussed along with the effects of ambient noise level on the frequency-based noise generated by wind turbines. The effects of wind turbine noise on television and radio signals are investigated with emphasis on quality of sound and video. The use of the blade element momentum (BEM) theorem to calculate critical wind turbine performance parameters is explained.

Estimated wind turbine procurement costs as a function of output rating and rotor blade diameter are provided. Sample calculations to predict the electricity cost per kilowatt hour as a function of turbine output power, operating cost, and maintenance cost are included. More elaborate methods of computing cost per kilowatt hour for wind turbine energy along with realistic assumptions are discussed. Procurement costs for critical components manufactured in Europe and the U.S. are summarized. Estimated costs, operating height, and mechanical strength requirements are briefly discussed for various tower types and configurations.

References

[1] S. Mertens, *Wind Energy in Built Environments*, 1989, Multiscience, Essex, U.K., p. 16.
[2] R.E. Wilson, S.N. Walker, and P.B. Lissaman, Aerodynamics of Darrieus rotor, *AIAA J. Aircraft*, 15, 1023, 1976.
[3] S. Mertens, *Wind Energy in Built Environments*, 1989, Multiscience, Essex, U.K., p. 79.
[4] A.J. Wortman, *Introduction to Wind Turbine Engineering*, 1983, Butterworth, Boston, p. 29.
[5] J.F. Walker and N. Jenkins, *Wind Energy Technology*, 2002, John Wiley & Sons, Chichester.
[6] O.L. Martin and J.Hansen, *Aerodynamics of Wind Turbines*, 2nd ed., 1992, James & James, London.

Index

Printed and bound by CPI Group (UK) Ltd, Croydon, CR0 4YY

21/10/2024

01777109-0003